CAMBRIDGE LIBRARY COLLECTION

Books of enduring scholarly value

Darwin

Two hundred years after his birth and 150 years after the publication of 'On the Origin of Species', Charles Darwin and his theories are still the focus of worldwide attention. This series offers not only works by Darwin, but also the writings of his mentors in Cambridge and elsewhere, and a survey of the impassioned scientific, philosophical and theological debates sparked by his 'dangerous idea'.

The Formation of Vegetable Mould through the Action of Worms

This book, published in 1881, was the result of many years of experimentation and observation by Darwin in the open-air laboratory of his garden at Down House in Kent. As he wrote in his introduction, the subject of soil disturbance by worms 'may appear an insignificant one, but we shall see that it possesses some interest'. He goes on to demonstrate the immensity – in size and over time – of the accumulated tiny movements of soil by earthworms, and their vital role in aerating the soil and breaking down vegetable material to keep the topsoil, the growing medium for all plant life and thus vital to human existence, fertile and healthy. At a time when there is huge interest in growing food organically and keeping soil in good condition without using artificial fertilisers, Darwin's insights are as important, and his descriptions of his experiments as fascinating, as they were in the late nineteenth century.

The Formation of Vegetable Mould through the Action of Worms

With Observations on their Habits

CHARLES DARWIN

CAMBRIDGE UNIVERSITY PRESS

CAMBRIDGE UNIVERSITY PRESS

Cambridge New York Melbourne Madrid Cape Town Singapore São Paolo Delhi

Published in the United States of America by Cambridge University Press, New York

www.cambridge.org
Information on this title: www.cambridge.org/9781108005128

This edition first published 1881
This digitally printed version 2009

ISBN 978-1-108-00512-8

THE FORMATION

OF

VEGETABLE MOULD,

THROUGH THE

ACTION OF WORMS,

WITH

OBSERVATIONS ON THEIR HABITS.

BY CHARLES DARWIN, LL.D., F.R.S.

WITH ILLUSTRATIONS

LONDON:
JOHN MURRAY, ALBEMARLE STREET.
1881.

CONTENTS.

CHAPTER I.

HABITS OF WORMS.

CHAPTER II.

HABITS OF WORMS—*continued.*

CHAPTER III.

THE AMOUNT OF FINE EARTH BROUGHT UP BY WORMS TO THE SURFACE.

CHAPTER IV.

THE PART WHICH WORMS HAVE PLAYED IN THE BURIAL OF ANCIENT BUILDINGS.

CHAPTER V.

THE ACTION OF WORMS IN THE DENUDATION OF THE LAND.

CHAPTER VI.

THE DENUDATION OF THE LAND—*continued.*

CHAPTER VII.

CONCLUSION.

THE

FORMATION OF VEGETABLE MOULD,

THROUGH THE ACTION OF WORMS, WITH OBSERVATIONS ON THEIR HABITS.

INTRODUCTION.

THE share which worms have taken in the formation of the layer of vegetable mould, which covers the whole surface of the land in every moderately humid country, is the subject of the present volume. This mould is generally of a blackish colour and a few inches in thickness. In different districts it differs but little in appearance, although it may rest on various subsoils. The uniform fineness of the particles of which it is composed is one of its chief characteristic features; and this may be well observed in any gravelly country, where a recently-ploughed field immediately adjoins one which has long remained undisturbed for pasture, and where the vegetable mould is exposed on the sides of a ditch or hole. The subject may appear an insignificant one, but we shall see that it possesses some interest; and the maxim "de minimis lex non curat," does not apply to science. Even Elie de Beaumont, who generally undervalues small agencies and their accumulated effects, remarks.* "la "couche très-mince de la terre végétale est un "monument d'une haute antiquité, et, par le "fait de sa permanence, un objet digne d'oc- "cuper le géologue, et capable de lui fournir "des remarques intéressantes." Although the superficial layer of vegetable mould as a whole no doubt is of the highest antiquity, yet in regard to its permanence, we shall hereafter see reason to believe that its component particles are in most cases removed at not a very slow rate, and are replaced by others due to the disintegration of the underlying materials.

As I was led to keep in my study during many months worms in pots filled with earth,

* 'Leçons de Géologie Pratique,' tom. i. 1845, p. 140.

I became interested in them, and wished to learn how far they acted consciously, and how much mental power they displayed. I was the more desirous to learn something on this head, as few observations of this kind have been made, as far as I know, on animals so low in the scale of organization and so poorly provided with sense-organs, as are earth-worms.

In the year 1837, a short paper was read by me before the Geological Society of London,* "On the Formation of Mould," in which it was shown that small fragments of burnt marl, cinders, &c., which had been thickly strewed over the surface of several meadows, were found after a few years lying at the depth of some inches beneath the turf, but still forming a layer. This apparent sinking of superficial bodies is due, as was first suggested to me by Mr. Wedgwood of Maer Hall in Staffordshire, to the large quantity of fine earth continually brought up to the surface by worms in the form of castings. These castings are sooner or later

* 'Transactions Geolog. Soc.' vol. v. p. 505. Read November 1, 1837.

spread out and cover up any object left on the surface. I was thus led to conclude that all the vegetable mould over the whole country has passed many times through, and will again pass many times through, the intestinal canals of worms. Hence the term "animal mould" would be in some respects more appropriate than that commonly used of "vegetable mould."

Ten years after the publication of my paper, M. D'Archiac, evidently influenced by the doctrines of Elie de Beaumont, wrote about my "singulière théorie," and objected that it could apply only to "les prairies basses et humides;" and that "les terres labourées, les bois, les prairies élevées, n'apportent aucune preuve à l'appui de cette manière de voir." * But M. D'Archiac must have thus argued from inner consciousness and not from observation, for worms abound to an extraordinary degree in kitchen gardens where the soil is continually worked, though in such loose soil they generally deposit their castings in any open cavities or within their old burrows instead of on the surface. Von Hensen estimates that there are

* 'Histoire des progrès de la Géologie,' tom. i. 1847, p. 224.

about twice as many worms in gardens as in corn-fields.* With respect to "prairies êlevées," I do not know how it may be in France, but nowhere in England have I seen the ground so thickly covered with castings as on commons, at a height of several hundred feet above the sea. In woods again, if the loose leaves in autumn are removed, the whole surface will be found strewed with castings. Dr. King, the superintendent of the Botanic Garden in Calcutta, to whose kindness I am indebted for many observations on earth-worms, informs me that he found, near Nancy in France, the bottom of the State forests covered over many acres with a spongy layer, composed of dead leaves and innumerable worm-castings. He there heard the Professor of "Aménagement des Forêts" lecturing to his pupils, and pointing out this case as a "beautiful example of the natural cultiva-"tion of the soil; for year after year the "thrown-up castings cover the dead leaves; "the result being a rich humus of great "thickness."

* 'Zeitschrift für wissenschaft. Zoologie,' B. xxviii. 1877, p. 361.

In the year 1869, Mr. Fish* rejected my conclusions with respect to the part which worms have played in the formation of vegetable mould, merely on account of their assumed incapacity to do so much work. He remarks that "considering their weakness and their "size, the work they are represented to "have accomplished is stupendous." Here we have an instance of that inability to sum up the effects of a continually recurrent cause, which has often retarded the progress of science, as formerly in the case of geology, and more recently in that of the principle of evolution.

Although these several objections seemed to me to have no weight, yet I resolved to make more observations of the same kind as those published, and to attack the problem on another side; namely, to weigh all the castings thrown up within a given time in a measured space, instead of ascertaining the rate at which objects left on the surface were buried by worms. But some of my observations have been rendered almost superfluous by an admirable paper by Von Hensen,

* 'Gardeners' Chronicle,' April 17, 1869, p. 418.

already alluded to, which appeared in 1877. Before entering on details with respect to the castings, it will be advisable to give some account of the habits of worms from my own observations and from those of other naturalists.

CHAPTER I.

HABITS OF WORMS.

Nature of the sites inhabited—Can live long under water—Nocturnal—Wander about at night—Often lie close to the mouths of their burrows, and are thus destroyed in large numbers by birds—Structure—Do not possess eyes, but can distinguish between light and darkness—Retreat rapidly when brightly illuminated, not by a reflex action—Power of attention—Sensitive to heat and cold—Completely deaf—Sensitive to vibrations and to touch—Feeble power of smell—Taste—Mental qualities—Nature of food—Omnivorous—Digestion—Leaves before being swallowed, moistened with a fluid of the nature of the pancreatic secretion—Extra-stomachal digestion—Calciferous glands, structure of—Calcareous concretions formed in the anterior pair of glands—The calcareous matter primarily an excretion, but secondarily serves to neutralise the acids generated during the digestive process.

EARTH-WORMS are distributed throughout the world under the form of a few genera, which externally are closely similar to one another. The British species of Lumbricus have never been carefully monographed; but we may judge of their probable number from those inhabiting neighbouring countries. In Scandinavia there are eight species, according to

Eisen;* but two of these rarely burrow in the ground, and one inhabits very wet places or even lives under the water. We are here concerned only with the kinds which bring up earth to the surface in the form of castings. Hoffmeister says that the species in Germany are not well known, but gives the same number as Eisen, together with some strongly marked varieties.†

Earth-worms abound in England in many different stations. Their castings may be seen in extraordinary numbers on commons and chalk-downs, so as almost to cover the whole surface, where the soil is poor and the grass short and thin. But they are almost or quite as numerous in some of the London parks, where the grass grows well and the soil appears rich. Even on the same field worms are much more frequent in some places than in others, without any visible difference in the nature of the soil. They abound in paved court-yards close to houses; and an instance will be given in which they had

* 'Bidrag till Skandinaviens Oligochætfauna,' 1871.

† 'Die bis jetzt bekannten Arten aus der Familie der Regenwürmer,' 1845.

burrowed through the floor of a very damp cellar. I have seen worms in black peat in a boggy field; but they are extremely rare, or quite absent in the drier, brown, fibrous peat, which is so much valued by gardeners. On dry, sandy or gravelly tracks, where heath with some gorse, ferns, coarse grass, moss and lichens alone grow, hardly any worms can be found. But in many parts of England, wherever a path crosses a heath, its surface becomes covered with a fine short sward. Whether this change of vegetation is due to the taller plants being killed by the occasional trampling of man and animals, or to the soil being occasionally manured by the droppings from animals, I do not know.* On such grassy paths worm-castings may often be seen. On a heath in Surrey, which was carefully examined, there were only a few castings on these paths, where they were much inclined;

* There is even some reason to believe that pressure is actually favourable to the growth of grasses, for Professor Buckman, who made many observations on their growth in the experimental gardens of the Royal Agricultural College, remarks ('Gardeners' Chronicle,' 1854, p. 619): "Another circumstance in the cultivation of grasses in the separate form or small patches, is the impossibility of rolling or treading them firmly, without which no pasture can continue good."

but on the more level parts, where a bed of fine earth had been washed down from the steeper parts and had accumulated to a thickness of a few inches, worm-castings abounded. These spots seemed to be overstocked with worms, so that they had been compelled to spread to a distance of a few feet from the grassy paths, and here their castings had been thrown up among the heath; but beyond this limit, not a single casting could be found. A layer, though a thin one, of fine earth, which probably long retains some moisture, is in all cases, as I believe, necessary for their existence; and the mere compression of the soil appears to be in some degree favourable to them, for they often abound in old gravel walks, and in foot-paths across fields.

Beneath large trees few castings can be found during certain seasons of the year, and this is apparently due to the moisture having been sucked out of the ground by the innumerable roots of the trees; for such places may be seen covered with castings after the heavy autumnal rains. Although most coppices and woods support many worms, yet in a forest of tall and ancient beech-trees in Knole

Park, where the ground beneath was bare of all vegetation, not a single casting could be found over wide spaces, even during the autumn. Nevertheless, castings were abundant on some grass-covered glades and indentations which penetrated this forest. On the mountains of North Wales and on the Alps, worms, as I have been informed, are in most places rare; and this may perhaps be due to the close proximity of the subjacent rocks, into which worms cannot burrow during the winter so as to escape being frozen. Dr. McIntosh, however, found worm-castings at a height of 1500 feet on Schiehallion in Scotland. They are numerous on some hills near Turin at from 2000 to 3000 feet above the sea, and at a great altitude on the Nilgiri Mountains in South India and on the Himalaya.

Earth-worms must be considered as terrestrial animals, though they are still in one sense semi-aquatic, like the other members of the great class of annelids to which they belong. M. Perrier found that their exposure to the dry air of a room for only a single night was fatal to them. On the other hand he kept several large worms alive for nearly four months, completely submerged in water.* During the summer when the ground is dry, they penetrate to a considerable depth and cease to work, as they do during the winter when the ground is frozen. Worms are nocturnal in their habits, and at night may be seen crawling about in large numbers, but usually with their tails still inserted in their burrows. By the expansion of this part of their bodies, and with the help of the short, slightly reflexed bristles, with which their bodies are armed, they hold so fast that they can seldom be dragged out of the ground without being torn into pieces.† During the day they remain in their burrows, except at the pairing season, when those which inhabit adjoining burrows expose the greater part of their bodies for an hour or two in the early morning. Sick

* I shall have occasion often to refer to M. Perrier's admirable memoir, 'Organisation des Lombriciens terrestres' in 'Archives de Zoolog. expér.' tom. iii. 1874, p. 372. C. F. Morren ('De Lumbrici terrestris,' 1829, p. 14) found that worms endured immersion for fifteen to twenty days in summer, but that in winter they died when thus treated.

† Morren, 'De Lumbrici terrestris,' &c., 1829, p. 67.

individuals, which are generally affected by the parasitic larvæ of a fly, must also be excepted, as they wander about during the day and die on the surface. After heavy rain succeeding dry weather, an astonishing number of dead worms may sometimes be seen lying on the ground. Mr. Galton informs me that on one such occasion (March, 1881), the dead worms averaged one for every two and a half paces in length on a walk in Hyde Park, four paces in width. He counted no less than 45 dead worms in one place in a length of sixteen paces. From the facts above given, it is not probable that these worms could have been drowned, and if they had been drowned they would have perished in their burrows. I believe that they were already sick, and that their deaths were merely hastened by the ground being flooded.

It has often been said that under ordinary circumstances healthy worms never, or very rarely, completely leave their burrows at night; but this is an error, as White of Selborne long ago knew. In the morning, after there has been heavy rain, the film of mud or of very fine sand over gravel-walks is often plainly marked with their tracks. I have noticed this from August to May, both months included, and it probably occurs during the two remaining months of the year when they are wet. On these occasions, very few dead worms could anywhere be seen. On January 31, 1881, after a long-continued and unusually severe frost with much snow, as soon as a thaw set in, the walks were marked with innumerable tracks. On one occasion, five tracks were counted crossing a space of only an inch square. They could sometimes be traced either to or from the mouths of the burrows in the gravel-walks, for distances between 2 or 3 up to 15 yards. I have never seen two tracks leading to the same burrow; nor is it likely, from what we shall presently see of their sense-organs, that a worm could find its way back to its burrow after having once left it. They apparently leave their burrows on a voyage of discovery, and thus they find new sites to inhabit.

Morren states * that worms often lie for hours almost motionless close beneath the mouths of their burrows. I have occasionally noticed the same fact with worms kept in

* 'De Lumbrici terrestris,' &c., p. 14.

pots in the house; so that by looking down into their burrows, their heads could just be seen. If the ejected earth or rubbish over the burrows be suddenly removed, the end of the worm's body may very often be seen rapidly retreating. This habit of lying near the surface leads to their destruction to an immense extent. Every morning during certain seasons of the year, the thrushes and blackbirds on all the lawns throughout the country draw out of their holes an astonishing number of worms; and this they could not do, unless they lay close to the surface. It is not probable that worms behave in this manner for the sake of breathing fresh air, for we have seen that they can live for a long time under water. I believe that they lie near the surface for the sake of warmth, especially in the morning; and we shall hereafter find that they often coat the mouths of their burrows with leaves, apparently to prevent their bodies from coming into close contact with the cold damp earth. It is said that they completely close their burrows during the winter.

Structure.—A few remarks must be made on this subject. The body of a large worm consists of from 100 to 200 almost cylindrical rings or segments, each furnished with minute bristles. The muscular system is well developed. Worms can crawl backwards as well as forwards, and by the aid of their affixed tails can retreat with extraordinary rapidity into their burrows. The mouth is situated at the anterior end of the body, and is provided with a little projection (lobe or lip, as it has been variously called) which is used for prehension. Internally, behind the mouth, there is a strong pharynx, shown in the accompanying diagram (Fig. 1) which is pushed forwards when the animal eats, and this part corresponds, according to Perrier, with the protrudable trunk or proboscis of other annelids. The pharynx leads into the œsophagus, on each side of which in the lower part there are three pairs of large glands, which secrete a surprising amount of carbonate of lime. These calciferous glands are highly remarkable, for nothing like them is known in any other animal. Their use will be discussed when we treat of the digestive process. In most of the species, the œsophagus is enlarged into a crop in front of the gizzard. This latter organ is lined with a smooth thick chitinous membrane, and is surrounded by weak longitudinal, but by powerful transverse muscles. Perrier saw these muscles in energetic action; and, as he remarks, the trituration of the food must be chiefly effected by this organ, for worms possess no jaws or teeth of any kind. Grains of sand and small stones, from the $\frac{1}{20}$ to a little more than the $\frac{1}{10}$ inch in diameter, may generally be found in their gizzards and intestines. As it is certain that worms swallow many little stones, independently of those swallowed while excavating their burrows, it is probable that they serve, like mill-stones, to triturate their food. The gizzard opens into the intestine,

Fig. 1.

Diagram of the alimentary canal of an earthworm (Lumbricus), copied from Ray Lankester in 'Quart. Journ. of Microscop. Soc.' vol. xv. N.S. pl. vii.

which runs in a straight course to the vent at the posterior end of the body. The intestine presents a remarkable structure, the typhosolis, or, as the old anatomists called it, an intestine within an intestine; and Claparède* has shown that this consists of a deep longitudinal involution of the walls of the intestine, by which means an extensive absorbent surface is gained.

The circulatory system is well developed. Worms breathe by their skin, as they do not possess any special respiratory organs. The two sexes are united in the same individual, but two individuals pair together. The nervous system is fairly well developed; and the two almost confluent cerebral ganglia are situated very near to the anterior end of the body.

Senses.—Worms are destitute of eyes, and at first I thought that they were quite insensible to light; for those kept in confinement were repeatedly observed by the aid of a candle, and others out of doors by the aid of a lantern, yet they were rarely alarmed, although extremely timid animals. Other

* Histolog. Untersuchungen über die Regenwürmer. 'Zeitschrift für wissenschaft. Zoologie,' B. xix., 1869, p. 611.

persons have found no difficulty in observing worms at night by the same means.*

Hoffmeister, however, states † that worms, with the exception of a few individuals, are extremely sensitive to light; but he admits that in most cases a certain time is requisite for its action. These statements led me to watch on many successive nights worms kept in pots, which were protected from currents of air by means of glass plates. The pots were approached very gently, in order that no vibration of the floor should be caused. When under these circumstances worms were illuminated by a bull's-eye lantern having slides of dark red and blue glass, which intercepted so much light that they could be seen only with some difficulty, they were not at all affected by this amount of light, however long they were exposed to it. The light, as far as I could judge, was brighter than that from the full moon. Its colour apparently made no difference in the result. When they were

* For instance, Mr. Bridgman and Mr. Newman ('The Zoologist,' vol. vii. 1849, p. 2576), and some friends who observed worms for me.

† 'Familie der Regenwürmer,' 1845, p. 18.

illuminated by a candle, or even by a bright paraffin lamp, they were not usually affected at first. Nor were they when the light was alternately admitted and shut off. Sometimes, however, they behaved very differently, for as soon as the light fell on them, they withdrew into their burrows with almost instantaneous rapidity. This occurred perhaps once out of a dozen times. When they did not withdraw instantly, they often raised the anterior tapering ends of their bodies from the ground, as if their attention was aroused or as if surprise was felt; or they moved their bodies from side to side as if feeling for some object. They appeared distressed by the light; but I doubt whether this was really the case, for on two occasions after withdrawing slowly, they remained for a long time with their anterior extremities protruding a little from the mouths of their burrows, in which position they were ready for instant and complete withdrawal.

When the light from a candle was concentrated by means of a large lens on the anterior extremity, they generally withdrew instantly; but this concentrated light failed to act perhaps once out of half a dozen trials. The light was on one occasion concentrated on a worm lying beneath water in a saucer, and it instantly withdrew into its burrow. In all cases the duration of the light, unless extremely feeble, made a great difference in the result; for worms left exposed before a paraffin lamp or a candle invariably retreated into their burrows within from five to fifteen minutes; and if in the evening the pots were illuminated before the worms had come out of their burrows, they failed to appear.

From the foregoing facts it is evident that light affects worms by its intensity and by its duration. It is only the anterior extremity of the body, where the cerebral ganglia lie, which is affected by light, as Hoffmeister asserts, and as I observed on many occasions. If this part is shaded, other parts of the body may be fully illuminated, and no effect will be produced. As these animals have no eyes, we must suppose that the light passes through their skins, and in some manner excites their cerebral ganglia. It appeared at first probable that the different manner in which they were affected on different occasions might be explained, either by the degree of extension of their skin and its consequent transparency, or by some particular incidence of the light; but I could discover no such relation. One thing was manifest, namely that when worms were employed in dragging leaves into their burrows or in eating them, and even during the short intervals whilst they rested from their work, they either did not perceive the light or were regardless of it; and this occurred even when the light was concentrated on them through a large lens. So, again, whilst they are paired, they will remain for an hour or two out of their burrows, fully exposed to the morning light; but it appears from what Hoffmeister says that a light will occasionally cause paired individuals to separate.

When a worm is suddenly illuminated and dashes like a rabbit into its burrow—to use the expression employed by a friend—we are at first led to look at the action as a reflex one. The irritation of the cerebral ganglia appears to cause certain muscles to contract in an inevitable manner, independently of the will or consciousness of the animal, as if it were an automaton. But the different effect which a light produced on different occasions, and especially the fact that a worm when in any way employed and in the intervals of such employment, whatever set of muscles and ganglia may then have been brought into play, is often regardless of light, are opposed to the view of the sudden withdrawal being a simple reflex action. With the higher animals, when close attention to some object leads to the disregard of the impressions which other objects must be producing on them, we attribute this to their attention being then absorbed; and attention implies the presence of a mind. Every sportsman knows that he can approach animals whilst they are grazing, fighting or courting, much more easily than at other times. The state, also, of the nervous system of the higher animals differs much at different times, for instance, a horse is much more readily startled at one time than at another. The comparison here implied between the actions of one of the higher animals and of one so low in the scale as an earth-worm, may appear far-fetched; for we thus attribute to the worm attention and some mental power, nevertheless I can see no reason to doubt the justice of the comparison.

Although worms cannot be said to possess the power of vision, their sensitiveness to light enables them to distinguish between day and night; and they thus escape extreme danger from the many diurnal animals which prey on them. Their withdrawal into their burrows during the day appears, however, to have become an habitual action; for worms kept in pots covered by glass-plates, over which sheets of black paper were spread, and placed before a north-east window, remained during the day-time in their burrows and came out every night; and they continued thus to act for a week. No doubt a little light may have entered between the sheets of glass and the blackened paper; but we know from the trials with coloured glass, that worms are indifferent to a small amount of light.

Worms appear to be less sensitive to moderate radiant heat than to a bright light. I judge of this from having held at different times a poker heated to dull redness near some worms, at a distance which caused a very sensible degree of warmth in my hand. One of them took no notice; a second withdrew into its burrow, but not quickly; the third and fourth much more quickly, and the fifth as quickly as possible. The light from a candle, concentrated by a lens and passing through a sheet of glass which would intercept most of the heat-rays, generally caused a much more rapid retreat than did the heated poker. Worms are sensitive to a low temperature, as may be inferred from their not coming out of their burrows during a frost.

Worms do not possess any sense of hearing. They took not the least notice of the shrill notes from a metal whistle, which was repeatedly sounded near them; nor did they of the deepest and loudest tones of a bassoon. They were indifferent to shouts, if care was taken that the breath did not strike them. When placed on a table close to the keys of a piano, which was played as loudly as possible, they remained perfectly quiet.

Although they are indifferent to undulations in the air audible by us, they are extremely sensitive to vibrations in any solid object. When the pots containing two worms which had remained quite indifferent to the sound of the piano, were placed on this instrument, and the note C in the bass clef was struck, both instantly retreated into their burrows. After a time they emerged, and when G above the line in the treble clef was struck they again retreated. Under similar circumstances on another night one worm dashed into its burrow on a very high note being struck only once, and the other worm when C in the treble clef was struck. On these occasions the worms were not touching the sides of the pots, which stood in saucers; so that the vibrations, before reaching their bodies, had to pass from the sounding board of the piano, through the saucer, the bottom of the pot and the damp, not very compact earth on which they lay with their tails in their burrows. They often showed their sensitiveness when the pot in which they lived, or the table on which the pot stood, was accidentally and lightly struck; but they appeared less sensitive to such jars than to the vibrations of the piano; and their sensitiveness to jars varied much at different times. It has often been said that if the ground is beaten or otherwise made to tremble, worms believe that they are pursued by a mole and leave their burrows. I beat the ground in many places where worms abounded, but not one emerged. When, however, the ground is dug with a fork and is violently disturbed beneath a worm, it will often crawl quickly out of its burrow.

The whole body of a worm is sensitive to contact. A slight puff of air from the mouth causes an instant retreat. The glass plates placed over the pots did not fit closely, and blowing through the very narrow chinks thus left, often sufficed to cause a rapid retreat. They sometimes perceived the eddies in the air caused by quickly removing the glass plates. When a worm first comes out of its burrow, it generally moves the much extended anterior extremity of its body from side to side in all directions, apparently as an organ of touch; and there is some reason to believe, as we shall see in the next chapter, that they are thus enabled to gain a general notion of the form of an object. Of all their senses that of touch, including in this term the perception of a vibration, seems much the most highly developed.

In worms the sense of smell apparently is confined to the perception of certain odours, and is feeble. They were quite indifferent to my breath, as long as I breathed on them very gently. This was tried, because it appeared possible that they might thus be warned of the approach of an enemy. They exhibited the same indifference to my breath whilst I chewed some tobacco, and while a pellet of cotton-wool with a few drops of mille-fleurs perfume or of acetic acid was kept in my mouth. Pellets of cotton-wool soaked in tobacco juice, and in millefleurs perfume, and in paraffin, were held with pincers and were waved about within two or three inches of several worms, but they took no notice. On one or two occasions, however, when acetic acid had been placed on the pellets, the worms appeared a little uneasy, and this was probably due to the irritation of their skins. The perception of such unnatural odours would be of no service to worms; and as such timid creatures would almost certainly exhibit some signs of any new impression, we may conclude that they did not perceive these odours.

The result was different when cabbage-leaves and pieces of onion were employed, both of which are devoured with much relish by worms. Small square pieces of fresh and half-decayed cabbage-leaves and of onion bulbs were on nine occasions buried in my pots, beneath about $\frac{1}{4}$ of an inch of common garden soil; and they were always discovered by the worms. One bit of cabbage was discovered and removed in the course of two hours; three were removed by the next morning, that is, after a single night; two others after two nights; and the seventh bit after three nights. Two pieces of onion were discovered and removed after three nights. Bits of fresh raw meat, of which worms are very fond, were buried, and were not discovered within forty-eight hours, during which time they had not become putrid. The earth above the various buried objects was generally pressed down only slightly, so as not to prevent the emission of any odour.

On two occasions, however, the surface was well watered, and was thus rendered somewhat compact. After the bits of cabbage and onion had been removed, I looked beneath them to see whether the worms had accidentally come up from below, but there was no sign of a burrow; and twice the buried objects were laid on pieces of tin-foil which were not in the least displaced. It is of course possible that the worms whilst moving about on the surface of the ground, with their tails affixed within their burrows, may have poked their heads into the places where the above objects were buried; but I have never seen worms acting in this manner. Some pieces of cabbage-leaf and of onion were twice buried beneath very fine ferruginous sand, which was slightly pressed down and well watered, so as to be rendered very compact, and these pieces were never discovered. On a third occasion the same kind of sand was neither pressed down nor watered, and the pieces of cabbage were discovered and removed after the second night. These several facts indicate that worms possess some power of smell; and that they discover by this means odoriferous and much-coveted kinds of food.

It may be presumed that all animals which feed on various substances possess the sense of taste, and this is certainly the case with worms. Cabbage-leaves are much liked by worms; and it appears that they can distinguish between different varieties; but this may perhaps be owing to differences in their texture. On eleven occasions pieces of the fresh leaves of a common green variety and of the red variety used for pickling were given them, and they preferred the green, the red being either wholly neglected or much less gnawed. On two other occasions, however, they seemed to prefer the red. Half-decayed leaves of the red variety and fresh leaves of the green were attacked about equally. When leaves of the cabbage, horse-radish (a favourite food) and of the onion were given together, the latter were always and manifestly preferred. Leaves of the cabbage, lime-tree, Ampelopis, parsnip (Pastinaca), and celery (Apium) were likewise given together; and those of the celery were first eaten. But when leaves of cabbage, turnip, beet, celery, wild cherry and carrots were given together, the two latter kinds, especially those of the carrot, were preferred to all the others, including those of celery. It was also manifest after many trials that wild cherry leaves were greatly preferred to those of the lime-tree and hazel (Corylus). According to Mr. Bridgman the half-decayed leaves of *Phlox verna* are particularly liked by worms.*

Pieces of the leaves of cabbage, turnip, horse-radish and onion were left on the pots during 22 days, and were all attacked and had to be renewed; but during the whole of this time leaves of an Artemisia and of the culinary sage, thyme and mint, mingled with the above leaves, were quite neglected excepting those of the mint, which were occasionally and very slightly nibbled. These latter four kinds of leaves do not differ in texture in a manner which could make them disagreeable to worms; they all have a strong taste, but so have the four first mentioned kinds of leaves; and the wide difference in the result must be attributed to a preference by the worms for one taste over another.

* 'The Zoologist,' vol. vii. 1849, p. 2576.

D

Mental Qualities.—There is little to be said on this head. We have seen that worms are timid. It may be doubted whether they suffer as much pain when injured, as they seem to express by their contortions. Judging by their eagerness for certain kinds of food, they must enjoy the pleasure of eating. Their sexual passion is strong enough to overcome for a time their dread of light. They perhaps have a trace of social feeling, for they are not disturbed by crawling over each other's bodies, and they sometimes lie in contact. According to Hoffmeister they pass the winter either singly or rolled up with others into a ball at the bottom of their burrows.* Although worms are so remarkably deficient in the several sense-organs, this does not necessarily preclude intelligence, as we know from such cases as those of Laura Bridgman; and we have seen that when their attention is engaged, they neglect impressions to which they would otherwise have attended; and attention indicates the presence of a mind of some kind. They are also much more easily excited at certain times than at others.

* 'Familie der Regenwürmer,' p. 13.

They perform a few actions instinctively, that is, all the individuals, including the young, perform such actions in nearly the same fashion. This is shown by the manner in which the species of Perichæta eject their castings, so as to construct towers; also by the manner in which the burrows of the common earth-worm are smoothly lined with fine earth and often with little stones, and the mouths of their burrows with leaves. One of their strongest instincts is the plugging up the mouths of their burrows with various objects; and very young worms act in this manner. But some degree of intelligence appears, as we shall see in the next chapter, to be exhibited in this work,—a result which has surprised me more than anything else in regard to worms.

Food and Digestion.—Worms are omnivorous. They swallow an enormous quantity of earth, out of which they extract any digestible matter which it may contain; but to this subject I must recur. They also consume a large number of half-decayed leaves of all kinds, excepting a few which have an unpleasant taste or are too tough for them;

likewise petioles, peduncles and decayed flowers. But they will also consume fresh leaves, as I have found by repeated trials. According to Morren* they will eat particles of sugar and liquorice; and the worms which I kept drew many bits of dry starch into their burrows, and a large bit had its angles well rounded by the fluid poured out of their mouths. But as they often drag particles of soft stone, such as of chalk, into their burrows, I feel some doubt whether the starch was used as food. Pieces of raw and roasted meat were fixed several times by long pins to the surface of the soil in my pots, and night after night the worms could be seen tugging at them, with the edges of the pieces engulfed in their mouths, so that much was consumed. Raw fat seems to be preferred even to raw meat or to any other substance which was given them, and much was consumed. They are cannibals, for the two halves of a dead worm placed in two of the pots were dragged into the burrows and gnawed; but as far as I could judge, they prefer fresh to putrid meat, and in so far I differ from Hoffmeister.

* 'De Lumbrici terrestris' p. 19.

Léon Frédéricq states* that the digestive fluid of worms is of the same nature as the pancreatic secretion of the higher animals; and this conclusion agrees perfectly with the kinds of food which worms consume. Pancreatic juice emulsifies fat, and we have just seen how greedily worms devour fat; it dissolves fibrin, and worms eat raw meat; it converts starch into grape-sugar with wonderful rapidity, and we shall presently show that the digestive fluid of worms acts on starch.† But they live chiefly on half-decayed leaves; and these would be useless to them unless they could digest the cellulose forming the cell-walls; for it is well known that all other nutritious substances are almost completely withdrawn from leaves, shortly before they fall off. It has, however, now been ascertained that cellulose, though very little or not at all attacked by the gastric secretion of the higher animals, is acted on by that from the pancreas.‡

* 'Archives de Zoologie expérimentale,' tom. vii. 1878, p. 394.

† On the action of the pancreatic ferment, see 'A Text-Book of Physiology,' by Michael Foster, 2nd edit. pp. 198–203. 1878.

‡ Schmulewitsch, 'Action des Sucs digestifs sur la Cellulose.' Bull. de l'Acad. Imp. de St. Pétersbourg, tom. xxv. p. 549. 1879.

The half-decayed or fresh leaves which worms intend to devour, are dragged into the mouths of their burrows to a depth of from one to three inches, and are then moistened with a secreted fluid. It has been assumed that this fluid serves to basten their decay; but a large number of leaves were twice pulled out of the burrows of worms and kept for many weeks in a very moist atmosphere under a bell-glass in my study; and the parts which had been moistened by the worms did not decay more quickly in any plain manner than the other parts. When fresh leaves were given in the evening to worms kept in confinement and examined early on the next morning, therefore not many hours after they had been dragged into the burrows, the fluid with which they were moistened, when tested with neutral litmus paper, showed an alkaline reaction. This was repeatedly found to be the case with celery, cabbage and turnip leaves. Parts of the same leaves which had not been moistened by the worms, were pounded with a few drops of distilled water, and the juice thus extracted was not alkaline. Some leaves, however, which had been drawn into burrows out of doors, at an unknown antecedent period, were tried, and though still moist, they rarely exhibited even a trace of alkaline reaction.

The fluid, with which the leaves are bathed, acts on them whilst they are fresh or nearly fresh, in a remarkable manner; for it quickly kills and discolours them. Thus the ends of a fresh carrot-leaf, which had been dragged into a burrow, were found after twelve hours of a dark brown tint. Leaves of celery, turnip, maple, elm, lime, thin leaves of ivy, and occasionally those of the cabbage were similarly acted on. The end of a leaf of *Triticum repens*, still attached to a growing plant, had been drawn into a burrow, and this part was dark brown and dead, whilst the rest of the leaf was fresh and green. Several leaves of lime and elm removed from burrows out of doors were found affected in different degrees. The first change appears to be that the veins become of a dull reddish-orange. The cells with chlorophyll next lose more or less completely their green colour, and their contents finally become brown. The parts thus affected often appeared almost black by reflected light; but when viewed as a transparent object under the microscope, minute specks of light were transmitted, and this was not the case with the unaffected parts of the same leaves. These effects, however, merely show that the secreted fluid is highly injurious or poisonous to leaves; for nearly the same effects were produced in from one to two days on various kinds of young leaves, not only by artificial pancreatic fluid, prepared with or without thymol, but quickly by a solution of thymol by itself. On one occasion leaves of Corylus were much discoloured by being kept for eighteen hours in pancreatic fluid, without any thymol. With young and tender leaves immersion in human saliva during rather warm weather, acted in the same manner as the pancreatic fluid, but not so quickly. The leaves in all these cases often became infiltrated with the fluid.

Large leaves from an ivy plant growing on a wall were so tough that they could not be gnawed by worms, but after four days they were affected in a peculiar manner by the secretion poured out of their mouths. The upper surfaces of the leaves, over which the worms had crawled, as was shown by the dirt left on them, were marked in sinuous lines, by either a continuous or broken chain of whitish and often star-shaped dots, about 2 mm. in diameter. The appearance thus presented was curiously like that of a leaf, into which the larva of some minute insect had burrowed. But my son Francis, after making and examining sections, could nowhere find that the cell-walls had been broken down or that the epidermis had been penetrated. When the section passed through the whitish dots, the grains of chlorophyll were seen to be more or less discoloured, and some of the palisade and mesophyll cells contained nothing but broken down granular matter. These effects must be attributed to the transudation of the secretion through the epidermis into the cells.

The secretion with which worms moisten leaves likewise acts on the starch granules within the cells. My son examined some leaves of the ash and many of the lime, which had fallen off the trees and had been partly dragged into worm-burrows. It is known that with fallen leaves the starch-grains are preserved in the guard-cells of the stomata. Now in several cases the starch had partially or wholly disappeared from these cells, in the parts which had been moistened by the secretion; while they were still well preserved in the other parts of the same leaves. Sometimes the starch was dissolved out of only one of the two guard-cells. The nucleus in one case had disappeared, together with the starch-granules. The mere burying of lime-leaves in damp earth for nine days did not cause the destruction of the starch-granules. On the other hand, the immersion of fresh lime and cherry leaves for eighteen hours in artificial pancreatic fluid, led to the dissolution of the starch-granules in the guard-cells as well as in the other cells.

From the secretion with which the leaves are moistened being alkaline, and from its acting both on the starch-granules and on the protoplasmic contents of the cells, we may infer that it resembles in nature not saliva,* but pancreatic secretion; and we know from Frédéricq that a secretion of this

* Claparède doubts whether saliva is secreted by worms: see 'Zeitschrift für wissenschaft. Zoologie,' B. xix. 1869, p. 601.

kind is found in the intestines of worms. As the leaves which are dragged into the burrows are often dry and shrivelled, it is indispensable for their disintegration by the unarmed mouths of worms that they should first be moistened and softened; and fresh leaves, however soft and tender they may be, are similarly treated, probably from habit. The result is that they are partially digested before they are taken into the alimentary canal. I am not aware of any other case of extra-stomachal digestion having been recorded. The boa-constrictor bathes its prey with saliva, but this is solely for lubricating it. Perhaps the nearest analogy may be found in such plants as Drosera and Dionæa; for here animal matter is digested and converted into peptone not within a stomach, but on the surfaces of the leaves.

Calciferous Glands.—These glands (see Fig. 1), judging from their size and from their rich supply of blood-vessels, must be of much importance to the animal. But almost as many theories have been advanced on their use as there have been observers. They consist of three pairs, which in the common earth-worm debouch into the alimentary canal in advance of the gizzard, but posteriorly to it in Urochtæa and some other genera.* The two posterior pairs are formed by lamellæ, which according to Claparède, are diverticula from the œsophagus.† These lamellæ are coated with a pulpy cellular layer, with the outer cells lying free in infinite numbers. If one of these glands is punctured and squeezed, a quantity of white pulpy matter exudes, consisting of these free cells. They are minute, and vary in diameter from 2 to 6 μ. They contain in their centres a little excessively fine granular matter; but they look so like oil globules that Claparède and others at first treated them with ether. This produces no effect; but they are quickly dissolved with effervescence in acetic acid, and when oxalate of ammonia is added to the solution a white precipitate is thrown down. We may therefore conclude that they contain carbonate of lime. If the cells

* Perrier, 'Archives de Zoolog. expér.' July, 1874, pp. 416, 419.

† 'Zeitschrift für wissenschaft. Zoologie,' B. xix. 1869, pp. 603–606.

are immersed in a very little acid, they become more transparent, look like ghosts, and are soon lost to view; but if much acid is added, they disappear instantly. After a very large number have been dissolved, a flocculent residue is left, which apparently consists of the delicate ruptured cell-walls. In the two posterior pairs of glands the carbonate of lime contained in the cells occasionally aggregates into small rhombic crystals or into concretions, which lie between the lamellæ; but I have seen only one, and Claparède only a very few such cases.

The two anterior glands differ a little in shape from the four posterior ones, by being more oval. They differ also conspicuously in generally containing several small, or two or three larger, or a single very large concretion of carbonate of lime, as much as 1½ mm. in diameter. When a gland includes only a few very small concretions, or, as sometimes happens, none at all, it is easily overlooked. The large concretions are round or oval, and exteriorly almost smooth. One was found which filled up not only the whole gland, as is often the case, but its neck; so that it resembled an olive-oil flask in shape. These concretions when broken are seen to be more or less crystalline in structure. How they escape from the gland is a marvel; but that they do escape is certain, for they are often found in the gizzard, intestines, and in the castings of worms, both with those kept in confinement and those in a state of nature.

Claparède says very little about the structure of the two anterior glands, and he supposes that the calcareous matter of which the concretions are formed is derived from the four posterior glands. But if an anterior gland which contains only small concretions is placed in acetic acid and afterwards dissected, or if sections are made of such a gland without being treated with acid, lamellæ like those in the posterior glands and coated with cellular matter could be plainly seen, together with a multitude of free calciferous cells readily soluble in acetic acid. When a gland is completely filled with a single large concretion, there are no free cells, as these have been all consumed in forming the concretion. But if such a concretion, or one of only moderately large size is dissolved in acid, much membranous matter is left, which appears to consist of the remains of the formerly active lamellæ. After the formation and expulsion of a large concretion, new lamellæ must be developed in some manner. In one section made by my son, the process had apparently commenced, although the gland contained two rather large concretions, for near the walls several cylindrical and oval pipes were intersected, which were lined with cellular matter and were quite filled with free calciferous cells. A great enlargement in one direction of several oval pipes would give rise to the lamellæ.

Besides the free calciferous cells in which no nucleus was visible, other and rather larger free cells were seen on three occasions; and these contained a distinct nucleus and nucleolus. They were only so far acted on by acetic acid that the nucleus was thus rendered more distinct. A very small concretion was removed from between two of the lamellæ within an anterior gland. It was embedded in pulpy cellular matter, with many free calciferous cells, together with a multitude of the larger, free, nucleated cells, and these latter cells were not acted on by acetic acid, while the former were dissolved. From this and other such cases I am led to suspect that the calciferous cells are developed from the larger nucleated ones; but how this is effected was not ascertained.

When an anterior gland contains several minute concretions, some of these are generally angular or crystalline in outline, while the greater number are rounded with an irregular mulberry-like surface. Calciferous cells adhered to many parts of these mulberry-like masses, and their gradual disappearance could be traced while they still remained attached. It was thus evident that the concretions are formed from the lime contained within the free calciferous cells. As the smaller concretions increase in size, they come into contact and unite, thus enclosing the now functionless lamellæ; and by such steps the formation of the largest concretions could be followed. Why the process regularly takes place in the two anterior glands, and only rarely in the four posterior glands is quite unknown. Morren says that these glands disappear during the winter; and I have seen some instances of this fact, and others in which either the anterior or posterior glands were at this season so shrunk and empty, that they could be distinguished only with much difficulty.

With respect to the function of the calciferous glands, it is probable that they primarily serve as organs of excretion, and secondarily as an aid to digestion. Worms consume many fallen leaves; and it is known that lime goes on accumulating in leaves until they drop off the parent-plant, instead of being re-absorbed into the stem or roots, like various other organic and inorganic substances.* The ashes of a leaf of an acacia have been known to contain as much as 72 per cent. of lime. Worms therefore would be liable to become charged with this earth, unless there were some special means for its excretion; and the calciferous glands are well adapted for this purpose. The worms which live in mould close over the chalk, often have their intestines filled with this substance, and their castings are almost white.

* De Vries, 'Landwirth. Jahrbücher,' 1881, p. 77.

Here it is evident that the supply of calcareous matter must be superabundant. Nevertheless with several worms collected on such a site, the calciferous glands contained as many free calciferous cells, and fully as many and large concretions, as did the glands of worms which lived where there was little or no lime; and this indicates that the lime is an excretion, and not a secretion poured into the alimentary canal for some special purpose.

On the other hand, the following considerations render it highly probable that the carbonate of lime, which is excreted by the glands, aids the digestive process under ordinary circumstances. Leaves during their decay generate an abundance of various kinds of acids, which have been grouped together under the term of humus acids. We shall have to recur to this subject in our fifth chapter, and I need here only say that these acids act strongly on carbonate of lime. The half-decayed leaves which are swallowed in such large quantities by worms would, therefore, after they have been moistened and triturated in the alimentary canal, be apt to produce such acids. And in the case of several worms, the contents of the alimentary canal were found to be plainly acid, as shown by litmus paper. This acidity cannot be attributed to the nature of the digestive fluid, for pancreatic fluid is alkaline; and we have seen that the secretion which is poured out of the mouths of worms for the sake of preparing the leaves for consumption, is likewise alkaline. The acidity can hardly be due to uric acid, as the contents of the upper part of the intestine were often acid. In one case the contents of the gizzard were slightly acid, those of the upper intestines being more plainly acid. In another case the contents of the pharynx were not acid, those of the gizzard doubtfully so, while those of the intestine were distinctly acid at a distance of 5 cm. below the gizzard. Even with the higher herbivorous and omnivorous animals, the contents of the large intestine are acid. "This, however, is not caused by any acid "secretion from the mucous membrane; the "reaction of the intestinal walls in the larger "as in the small intestine is alkaline. It "must therefore arise from acid fermentations "going on in the contents themselves. . . . "In Carnivora the contents of the coecum "are said to be alkaline, and naturally the "amount of fermentation will depend largely "on the nature of the food."*

With worms not only the contents of the intestines, but their ejected matter or the castings, are generally acid. Thirty castings from different places were tested, and with three or four exceptions were found to be acid; and the exceptions may have been due to such castings not having been recently ejected; for some which were at first acid, were on the following morning, after being dried and again moistened, no longer acid; and this probably resulted from the humus acids being, as is known to be the case, easily decomposed. Five fresh castings from worms which lived in mould close over the chalk, were of a whitish colour and abounded with calcareous matter; and these were not in the least acid. This shows how effectually carbonate of lime neutralises the intestinal acids. When worms were kept in pots filled

* M. Foster, 'A Text-Book of Physiology,' 2nd edit. 1878, p. 243.

with fine ferruginous sand, it was manifest that the oxide of iron, with which the grains of silex were coated, had been dissolved and removed from them in the castings.

The digestive fluid of worms resembles in its action, as already stated, the pancreatic secretion of the higher animals; and in these latter, "pancreatic digestion is essentially "alkaline; the action will not take place "unless some alkali be present; and the "activity of an alkaline juice is arrested by "acidification, and hindered by neutraliza-"tion."* Therefore it seems highly probable that the innumerable calciferous cells, which are poured from the four posterior glands into the alimentary canal of worms, serve to neutralise more or less completely the acids there generated by the half-decayed leaves. We have seen that these cells are instantly dissolved by a small quantity of acetic acid, and as they do not always suffice to neutralise the contents of even the upper part of the alimentary canal, the lime is perhaps aggregated into concretions in the anterior pair of glands, in order that some may be

* M. Foster, *Ibid.* p. 200.

carried down to the posterior parts of the intestine, where these concretions would be rolled about amongst the acid contents. The concretions found in the intestines and in the castings often have a worn appearance, but whether this is due to some amount of attrition or of chemical corrosion could not be told. Claparède believes that they are formed for the sake of acting as mill-stones, and of thus aiding in the trituration of the food. They may give some aid in this way; but I fully agree with Perrier that this must be of quite subordinate importance, seeing that the object is already attained by stones being generally present in the gizzards and intestines of worms.

CHAPTER II.

HABITS OF WORMS—*continued.*

Manner in which worms seize objects—Their power of suction—The instinct of plugging up the mouths of their burrows—Stones piled over the burrows—The advantages thus gained—Intelligence shown by worms in their manner of plugging up their burrows—Various kinds of leaves and other objects thus used—Triangles of paper—Summary of reasons for believing that worms exhibit some intelligence—Means by which they excavate their burrows, by pushing away the earth and swallowing it—Earth also swallowed for the nutritious matter which it contains—Depth to which worms burrow, and the construction of their burrows—Burrows lined with castings, and in the upper part with leaves—The lowest part paved with little stones or seeds—Manner in which the castings are ejected—The collapse of old burrows—Distribution of worms—Tower-like castings in Bengal—Gigantic castings on the Nilgiri Mountains—Castings ejected in all countries.

IN the pots in which worms were kept, leaves were pinned down to the soil, and at night the manner in which they were seized could be observed. The worms always endeavoured to drag the leaves towards their burrows; and they tore or sucked off small fragments, whenever the leaves were sufficiently tender. They generally seized the thin edge of a leaf with their mouths, between the projecting upper and lower lip; the thick and strong pharynx being at the same time, as Perrier remarks, pushed forward within their bodies, so as to afford a point of resistance for the upper lip. In the case of broad flat objects they acted in a wholly different manner. The pointed anterior extremity of the body, after being brought into contact with an object of this kind, was drawn within the adjoining rings, so that it appeared truncated and became as thick as the rest of the body. This part could then be seen to swell a little; and this, I believe, is due to the pharynx being pushed a little forwards. Then by a slight withdrawal of the pharynx or by its expansion, a vacuum was produced beneath the truncated slimy end of the body whilst in contact with the object; and by this means the two adhered firmly together.* That under these circumstances a vacuum was produced was plainly

* Claparède remarks ('Zeitschrift für wissenschaft. Zoolog.' B. 19, 1869, p. 602) that the pharynx appears from its structure to be adapted for suction.

seen on one occasion, when a large worm lying beneath a flaccid cabbage leaf tried to drag it away; for the surface of the leaf directly over the end of the worm's body became deeply pitted. On another occasion a worm suddenly lost its hold on a flat leaf; and the anterior end of the body was momentarily seen to be cup-formed. Worms can attach themselves to an object beneath water in the same manner; and I saw one thus dragging away a submerged slice of an onion-bulb.

The edges of fresh or nearly fresh leaves affixed to the ground were often nibbled by the worms; and sometimes the epidermis and all the parenchyma on one side was gnawed completely away over a considerable space; the epidermis alone on the opposite side being left quite clean. The veins were never touched, and leaves were thus sometimes partly converted into skeletons. As worms have no teeth and as their mouths consist of very soft tissue, it may be presumed that they consume by means of suction the edges and the parenchyma of fresh leaves, after they have been softened by the digestive fluid. They cannot attack such strong leaves as those of sea-kale or large and thick leaves of ivy; though one of the latter after it had become rotten was reduced in parts to the state of a skeleton.

Worms seize leaves and other objects, not only to serve as food, but for plugging up the mouths of their burrows; and this is one of their strongest instincts. Leaves and petioles of many kinds, some flower-peduncles, often decayed twigs of trees, bits of paper, feathers, tufts of wool and horse-hairs are dragged into their burrows for this purpose. I have seen as many as seventeen petioles of a Clematis projecting from the mouth of one burrow, and ten from the mouth of another. Some of these objects, such as the petioles just named, feathers, &c., are never gnawed by worms. In a gravel walk in my garden I found many hundred leaves of a pine-tree (*P. austriaca* or *nigricans*) drawn by their bases into burrows. The surfaces by which these leaves are articulated to the branches are shaped in as peculiar a manner as is the joint between the leg-bones of a quadruped; and if these surfaces had been in the least gnawed, the fact would have been immediately visible, but there was no trace of gnawing. Of ordinary dicotyledonous leaves, all those which are dragged into burrows are not gnawed. I have seen as many as nine leaves of the lime-tree drawn into the same burrow, and not nearly all of them had been gnawed; but such leaves may serve as a store for future consumption. Where fallen leaves are abundant, many more are sometimes collected over the mouth of a burrow than can be used, so that a small pile of unused leaves is left like a roof over those which have been partly dragged in.

A leaf in being dragged a little way into a cylindrical burrow is necessarily much folded or crumpled. When another leaf is drawn in, this is done exteriorly to the first one, and so on with the succeeding leaves; and finally all become closely folded and pressed together. Sometimes the worm enlarges the mouth of its burrow, or makes a fresh one close by, so as to draw in a still larger number of leaves. They often or generally fill up the interstices between the drawn-in leaves with moist viscid earth ejected from their bodies; and thus the mouths of the burrows are securely plugged. Hundreds of such plugged burrows may be seen in many places, especially during the autumnal and early winter months. But, as will hereafter be shown, leaves are dragged into the burrows not only for plugging them up and for food, but for the sake of lining the upper part or mouth.

When worms cannot obtain leaves, petioles, sticks, &c., with which to plug up the mouths of their burrows, they often protect them by little heaps of stones; and such heaps of smooth rounded pebbles may frequently be seen on gravel-walks. Here there can be no question about food. A lady, who was interested in the habits of worms, removed the little heaps of stones from the mouths of several burrows and cleared the surface of the ground for some inches all round. She went out on the following night with a lantern, and saw the worms with their tails fixed in their burrows, dragging the stones inwards by the aid of their mouths, no doubt by suction. "After two nights some of the "holes had 8 or 9 small stones over "them; after four nights one had about "30, and another 34 stones."* One stone which had been dragged over the gravel-walk to the mouth of a burrow weighed two ounces; and this proves how strong worms are. But they show greater strength in sometimes displacing stones in a well-trodden gravel-walk; that they do so, may be inferred from the cavities left by the displaced stones being exactly filled by those lying over the mouths of adjoining burrows, as I have myself observed.

Work of this kind is usually performed during the night; but I have occasionally known objects to be drawn into the burrows during the day. What advantage the worms derive from plugging up the mouths of their burrows with leaves, &c., or from piling stones over them, is doubtful. They do not act in this manner at the times when they eject much earth from their burrows; for their castings then serve to cover the mouth. When gardeners wish to kill worms on a

* An account of her observations is given in the 'Gardeners Chronicle,' March 28th, 1868, p. 324.

lawn, it is necessary first to brush or rake away the castings from the surface, in order that the lime-water may enter the burrows.* It might be inferred from this fact that the mouths are plugged up with leaves, &c., to prevent the entrance of water during heavy rain; but it may be urged against this view that a few, loose, well-rounded stones are ill-adapted to keep out water. I have moreover seen many burrows in the perpendicularly cut turf-edgings to gravel-walks, into which water could hardly flow, as well plugged as burrows on a level surface. Can the plugs or piles of stones aid in concealing the burrows from scolopenders, which, according to Hoffmeister, † are the bitterest enemies of worms? Or may not worms when thus protected be able to remain with safety with their heads close to the mouths of their burrows, which we know that they like to do, but which costs so many of them their lives? Or may not the plugs check the free ingress, of the lowest stratum of air, when chilled by

* London's 'Gard. Mag.' xvii. p. 216, as quoted in the 'Catalogue of the British Museum Worms,' 1865, p. 327.

† 'Familie der Regenwürmer,' p. 19.

radiation at night, from the surrounding ground and herbage. I am inclined to believe in this latter view; firstly, because when worms were kept in pots in a room with a fire, in which case cold air could not enter the burrows, they plugged them up in a slovenly manner; and secondarily, because they often coat the upper part of their burrows with leaves, apparently to prevent their bodies from coming into close contact with the cold damp earth. But the plugging-up process may perhaps serve for all the above purposes.

Whatever the motive may be, it appears that worms much dislike leaving the mouths of their burrows open. Nevertheless they will reopen them at night, whether or not they can afterwards close them. Numerous open burrows may be seen on recently-dug ground, for in this case the worms eject their castings in cavities left in the ground, or in the old burrows, instead of piling them over the mouths of their burrows, and they cannot collect objects on the surface by which the mouths might be protected. So again on a recently disinterred pavement of a Roman villa at Abinger (hereafter to be described) the worms pertinaciously opened their burrows almost every night, when these had been closed by being trampled on, although they were rarely able to find a few minute stones wherewith to protect them.

Intelligence shown by worms in their manner of plugging up their burrows.—If a man had to plug up a small cylindrical hole, with such objects as leaves, petioles or twigs, he would drag or push them in by their pointed ends; but if these objects were very thin relatively to the size of the hole, he would probably insert some by their thicker or broader ends. The guide in his case would be intelligence. It seemed therefore worth while to observe carefully how worms dragged leaves into their burrows; whether by their tips or bases or middle parts. It seemed more especially desirable to do this in the case of plants not natives to our country; for although the habit of dragging leaves into their burrows is undoubtedly instinctive with worms, yet instinct could not tell them how to act in the case of leaves about which their progenitors knew nothing. If, moreover, worms acted solely through instinct or an unvarying inherited impulse, they would draw all kinds of leaves into their burrows in the same manner. If they have no such definite instinct, we might expect that chance would determine whether the tip, base or middle was seized. If both these alternatives are excluded, intelligence alone is left; unless the worm in each case first tries many different methods, and follows that alone which proves possible or the most easy; but to act in this manner and to try different methods makes a near approach to intelligence.

In the first place 227 withered leaves of various kinds, mostly of English plants, were pulled out of worm-burrows in several places. Of these, 181 had been drawn into the burrows by or near their tips, so that the foot-stalk projected nearly upright from the mouth of the burrow; 20 had been drawn in by their bases, and in this case the tips projected from the burrows; and 26 had been seized near the middle, so that these had been drawn in transversely and were much crumpled. Therefore 80 per cent. (always using the nearest whole number) had been drawn in by the tip, 9 per cent. by the base or footstalk, and 11 per cent. transversely or by the middle. This alone is almost sufficient to show that chance does not determine the manner in which leaves are dragged into the burrows.

Of the above 227 leaves, 70 consisted of the fallen leaves of the common lime-tree, which is almost certainly not a native of England. These leaves are much acuminated towards the tip, and are very broad at the base with a well-developed foot-stalk. They are thin and quite flexible when half-withered. Of the 70, 79 per cent. had been drawn in by or near the tip; 4 per cent. by or near the base; and 17 per cent. transversely or by the middle. These proportions agree very closely, as far as the tip is concerned, with those before given. But the percentage drawn in by the base is smaller, which may be attributed to the breadth of the basal part of the blade. We here, also, see that the presence of a foot-stalk, which it might have been expected would have tempted the worms as a convenient handle, has little or no influence in determining the manner in which lime leaves are dragged into the burrows.

The considerable proportion, viz., 17 per cent., drawn in more or less transversely depends no doubt on the flexibility of these half-decayed leaves. The fact of so many having been drawn in by the middle, and of some few having been drawn in by the base, renders it improbable that the worms first tried to draw in most of the leaves by one or both of these methods, and that they afterwards drew in 79 per cent. by their tips; for it is clear that they would not have failed in drawing them in by the base or middle.

The leaves of a foreign plant were next searched for, the blades of which were not more pointed towards the apex than towards the base. This proved to be the case with those of a laburnum (a hybrid between *Cytisus alpinus* and *laburnum*) for on doubling the terminal over the basal half, they generally fitted exactly; and when there was any difference, the basal half was a little the narrower. It might, therefore, have been expected that an almost equal number of these leaves would have been drawn in by the tip and base, or a slight excess in favour of the latter. But of 73 leaves (not included in the first lot of 227) pulled out of worm-burrows, 63 per cent. had been drawn in by the tip; 27 per cent. by the base, and 10 per cent. transversely. We here see that a far larger proportion, viz., 27 per cent. were drawn in by the base than in the case of lime leaves, the blades of which are very broad at the base, and of which only 4 per cent. had thus been drawn in. We may perhaps account for the fact of a still larger proportion of the laburnum leaves not having been drawn in by the base, by worms having acquired the habit of generally drawing in leaves by their tips and thus avoiding the foot-stalk. For the basal margin of the blade in many kinds of leaves forms a large angle with the foot-stalk; and if such a leaf were drawn in by the foot-stalk, the basal margin would come abruptly into contact with the ground on each side of the burrow, and would render the drawing in of the leaf very difficult.

Nevertheless worms break through their habit of avoiding the footstalk, if this part offers them the most convenient means for drawing leaves into their burrows. The leaves of the endless hybridised varieties of the Rhododendron vary much in shape; some are narrowest towards the base and others towards the apex. After they have fallen off, the blade on each side of the midrib often becomes curled up while drying, sometimes along the whole length, sometimes chiefly at the base, sometimes towards the apex. Out of 28 fallen leaves on one bed of peat in my garden, no less than 23 were narrower in the basal quarter than in the terminal quarter of their length; and this narrowness was chiefly due to the curling in of the margins. Out of 36 fallen leaves on another bed, in which different varieties of the Rhododendron grew, only 17 were narrower towards the base than towards the apex. My son William, who first called my attention to this case, picked up 237 fallen leaves in his garden (where the Rhododendron grows in the natural soil) and of these 65 per cent. could have been drawn by worms into their burrows more easily by the base or foot-stalk than by the tip; and this was partly due to the shape of the leaf and in a less degree to the curling in of the margins: 27 per cent. could have been drawn in more easily by the tip than by the base: and 8 per cent. with about equal ease by either end. The shape of a fallen leaf ought to be judged of before one end has been drawn into a burrow, for after this has happened, the free end, whether it be the base or apex, will dry more quickly than the end embedded in the damp ground; and the exposed margins of the free end will consequently tend to become more curled inwards than they were when the leaf was first seized by the worm. My son found 91 leaves which had been dragged by worms into their burrows, though not to a great depth; of these 66 per cent. had been drawn in by the base or foot-stalk; and 34 per cent. by the tip. In this case, therefore, the worms judged with a considerable degree of correctness how best to draw the withered leaves of this foreign plant into their burrows; notwithstanding that they had to depart from their usual habit of avoiding the foot-stalk.

On the gravel-walks in my garden a very large number of leaves of three species of Pinus (*P. austriaca*, *nigricans* and *sylvestris*) are regularly drawn into the mouths of worm-burrows. These leaves consist of two needles, which are of considerable length in the two first and short in the last named species, and are united to a common base; and it is by this part that they are almost invariably drawn into the burrows. I have seen only two or at most three exceptions to this rule with worms in a state of nature. As the sharply pointed needles diverge a little, and as several leaves are drawn into the same burrow, each tuft forms a perfect *chevaux de frise*. On two occasions many of these tufts were pulled up in the evening, but by the following morning fresh leaves had been pulled in, and the burrows were again well protected. These leaves could not be dragged into the burrows to any depth, except by their bases, as a worm cannot seize hold of the two needles at the same time, and if one alone were seized by the apex, the other would be pressed against the ground and would resist the entry of the seized one. This was manifest in the above mentioned two or three exceptional cases. In order, therefore that worms should do their work well, they must drag pine-leaves into their burrows by their bases, where the two needles are conjoined. But how they are guided in this work is a perplexing question.

This difficulty led my son Francis and myself to observe worms in confinement during several nights by the aid of a dim light, while they dragged the leaves of the above named pines into their burrows. They moved the anterior extremities of their bodies about the leaves, and on several occasions when they touched the sharp end of a needle they withdrew suddenly as if pricked. But I doubt whether they were hurt, for they are indifferent to very sharp objects, and will swallow even rose-thorns and small splinters of glass. It may also be doubted, whether the sharp ends of the needles serve to tell them that this is the wrong end to seize; for the points were cut off many leaves for a length of about one inch, and fifty-seven of them thus treated were drawn into the burrows by their bases, and not one by the cut-off ends. The worms in confinement often seized the needles near the middle and drew them towards the mouths of their burrows; and one worm tried in a senseless manner to drag them into the burrow by bending them. They sometimes collected many more leaves over the mouths of their burrows (as in the case formerly mentioned of lime-leaves) than could enter them. On other occasions, however, they behaved very differently; for as soon as they touched the base of a pine-leaf, this was seized, being sometimes completely engulfed in their mouths, or a point very near the base was seized, and the leaf was then quickly dragged or rather jerked into their burrows. It appeared both to my son and myself as if the worms instantly perceived as soon as they had seized a leaf in the proper manner. Nine such cases were observed, but in one of them the worm failed to drag the leaf into its burrow, as it was entangled by other leaves lying near. In another case a leaf stood nearly upright with the points of the needles partly inserted into a burrow, but how placed there was not seen; and then the worm reared itself up and seized the base, which was dragged into the mouth of the burrow by bowing the whole leaf. On the other hand, after a worm had seized the base of a leaf, this was on two occasions relinquished from some unknown motive.

As already remarked, the habit of plugging up the mouths of the burrows with various objects, is no doubt instinctive in worms; and a very young one, born in one of my pots, dragged for some little distance a Scotch-fir leaf, one needle of which was as long and almost as thick as its own body. No species of pine is endemic in this part of England, it is therefore incredible that the proper manner of dragging pine-leaves into the burrows can be instinctive with our worms. But as the worms on which the above observations were made, were dug up beneath or near some pines, which had been planted there about forty years, it was desirable to prove that their actions were not instinctive. Accordingly, pine-leaves were scattered on the ground in places far removed from any pine-tree, and 90 of them were drawn into the burrows by their bases. Only two were drawn in by the tips of the needles, and these were not real exceptions, as one was drawn in for a very short distance, and the two needles of the other cohered. Other pine-leaves were given to worms kept in pots in a warm room, and here the result was different; for out of 42 leaves drawn into the burrows, no less than 16 were drawn in by the tips of the needles. These worms, however, worked in a careless or slovenly manner; for the leaves were often drawn in to only a small depth; sometimes they were merely heaped over the mouths of the burrows, and sometimes none were drawn in. I believe that this carelessness may be accounted for by the air of the room being warm, and the worms consequently not being anxious to plug up their holes effectually. Pots tenanted by worms and covered with a net which allowed the entrance of cold air, were left out of doors for several nights, and now 72 leaves were all properly drawn in by their bases.

It might perhaps be inferred from the facts as yet given, that worms somehow gain a general notion of the shape or structure of pine leaves, and perceive that it is necessary for them to seize the base where the two needles are conjoined. But the following cases make this more than doubtful. The tips of a large number of needles of *P. austriaca* were cemented together with shell-lac dissolved in alcohol, and were kept for some days, until, as I believe, all odour or taste had been lost; and they were then scattered on the ground where no pine-trees grew, near burrows from which the plugging had been removed. Such leaves could have been drawn into the burrows with equal ease by either end; and judging from analogy and more especially from the case presently to be given of the petioles of *Clematis montana*, I expected that the apex would have been preferred. But the result was that out of 121 leaves with the tips cemented, which were drawn into burrows, 108 were drawn in by their bases, and only 13 by their tips. Thinking that the worms might possibly perceive and dislike the smell or taste of the shell-lac, though this was very improbable, especially after the leaves had been left out during several nights, the tips of the needles of many leaves were tied together with fine thread. Of leaves thus treated 150 were drawn into burrows—123 by the base and 27 by the tied tips; so that between four and five times as many were drawn in by the base as by the tip. It is possible that the short cut-off ends of the thread with which they were tied, may have tempted the worms to drag in a larger proportional number by the tips than when cement was used. Of the leaves with tied and cemented tips taken together (271 in number) 85 per cent. were drawn in by the base and 15 per cent. by the tips. We may therefore infer that it is not the divergence of the two needles which leads worms in a state of nature almost invariably to drag pine-leaves into their burrows by the base. Nor can it be the sharpness of the points of the needles which determines them; for, as we have seen, many leaves with the points cut off were drawn in by their bases. We are thus led to conclude, that with pine-leaves there must be something attractive to worms in the base, notwithstanding that few ordinary leaves are drawn in by the base or footstalk.

Petioles.—We will now turn to the petioles or foot-stalks of compound leaves, after the leaflets have fallen off. Those from *Clematis montana*, which grew over a verandah, were dragged early in January in large numbers into the burrows on an adjoining gravel-walk, lawn, and flower-bed. These petioles vary from $2\frac{1}{2}$ to $4\frac{1}{2}$ inches in length, are rigid and of nearly uniform thickness, except close to the base where they thicken rather abruptly, being here about twice as thick as in any other part. The apex is somewhat pointed, but soon withers and is then easily broken off. Of these petioles, 314 were pulled out of burrows in the above specified sites; and it was found that 76 per cent. had been drawn in by their tips, and 24 per cent. by their bases; so that those drawn in by the tip were a little more than thrice as many as those drawn in by the base. Some of those extracted from the well-beaten gravel-walk were kept separate from the others; and of these (59 in number) nearly five times as many had been drawn in by the tip as by the base; whereas of those extracted from the lawn and flower-bed, where from the soil yielding more easily, less care would be necessary in plugging up the burrows, the proportion of those drawn in by the tip (130) to those drawn in by the base (48) was rather less than three to one. That these petioles had been dragged into the burrows for plugging them up, and not for food, was manifest, as neither end, as far as I could see, had been gnawed. As several petioles are used to plug up the same burrow, in one case as many as 10, and in another case as many as 15, the worms may perhaps at first draw in a few by the thicker end so as to save labour; but afterwards a large majority are drawn in by the pointed end, in order to plug up the hole securely.

The fallen petioles of our native ash-tree were next observed, and the rule with most objects, viz., that a large majority are dragged into the burrows by the more pointed end, had not here been followed; and this fact much surprised me at first. These petioles vary in length from 5 to 8½ inches; they are thick and fleshy towards the base, whence they taper gently towards the apex, which is a little enlarged and truncated where the terminal leaflet had been originally attached. Under some ash-trees growing in a grass-field, 229 petioles were pulled out of worm burrows early in January, and of these 51·5 per cent. had been drawn in by the base, and 48·5 per cent. by the apex. This anomaly was however readily explained as soon as the thick basal part was examined; for in 78 out of 103 petioles, this part had been gnawed by worms, just above the horse-shoe shaped articulation. In most cases there could be no mistake about the gnawing; for ungnawed petioles which were examined after being exposed to the weather for eight additional weeks had not become more disintegrated or decayed near the base than elsewhere. It is thus evident that the thick basal end of the petiole is drawn in not solely for the sake of plugging up the mouths of the burrows, but as food. Even the narrow truncated tips of some few petioles had been gnawed; and this was the case in 6 out of 37 which were examined for this purpose. Worms, after having drawn in and gnawed the basal end, often push the petioles out of their burrows; and then drag in fresh ones, either by the base for food, or by the apex for plugging up the mouth more effectually. Thus, out of 37 petioles inserted by their tips, 5 had been previously drawn in by the base, for this part had been gnawed. Again,

I collected a handful of petioles lying loose on the ground close to some plugged-up burrows, where the surface was thickly strewed with other petioles which apparently had never been touched by worms; and 14 out of 47 (i.e. nearly one-third), after having had their bases gnawed had been pushed out of the burrows and were now lying on the ground. From these several facts we may conclude that worms draw in some petioles of the ash by the base to serve as food, and others by the tip to plug up the mouths of their burrows in the most efficient manner.

The petioles of *Robinia pseudo-acacia* vary from 4 or 5 to nearly 12 inches in length; they are thick close to the base before the softer parts have rotted off, and taper much towards the upper end. They are so flexible that I have seen some few doubled up and thus drawn into the burrows of worms. Unfortunately these petioles were not examined until February, by which time the softer parts had completely rotted off, so that it was impossible to ascertain whether worms had gnawed the bases, though this is in itself probable. Out of 121 petioles extracted from burrows early in February, 68 were embedded by the base, and 53 by the apex. On February 5 all the petioles which had been drawn into the burrows beneath a Robinia, were pulled up; and after an interval of eleven days, 35 petioles had been again dragged in, 19 by the base, and 16 by the apex. Taking these two lots together, 56 per cent. were drawn in by the base, and 44 per cent. by the apex. As all the softer parts had long ago rotted off, we may feel sure, especially in the latter case, that none had been drawn in as food. At this season, therefore, worms drag these petioles into their burrows indifferently by either end, a slight preference being given to the base. This latter fact may be accounted for by the difficulty of plugging up a burrow with objects so extremely thin as are the upper ends. In support of this view, it may be stated that out of the 16 petioles which had been drawn in by their upper ends, the more attenuated terminal portion of 7 had been previously broken off by some accident.

Triangles of paper.—Elongated triangles

were cut out of moderately stiff writing-paper, which was rubbed with raw fat on both sides, so as to prevent their becoming excessively limp when exposed at night to rain and dew. The sides of all the triangles were three inches in length, with the bases of 120 one inch, and of the other 183 half an inch in length. These latter triangles were very narrow or much acuminated.* As a check on the observations presently to be given, similar triangles in a damp state were seized by a very narrow pair of pincers at different points and at all inclinations with reference to the margins, and were then drawn into a short tube of the diameter of a worm-burrow. If seized by the apex, the triangle was drawn straight into the tube, with its margins infolded; if seized at some little distance from the apex, for instance at half an inch, this much was doubled back within the tube. So it was with the base and basal angles, though in this case the triangles offered, as might have been expected, much

* In these narrow triangles the apical angle is 9° 34′, and the basal angles 85° 13′. In the broader triangles the apical angle is 19° 10′ and the basal angles 80° 25′.

more resistance to being drawn in. If seized near the middle the triangle was doubled up, with the apex and base left sticking out of the tube. As the sides of the triangles were three inches in length, the result of their being drawn into a tube or into a burrow in different ways, may be conveniently divided into three groups: those drawn in by the apex or within an inch of it; those drawn in by the base or within an inch of it; and those drawn in by any point in the middle inch.

In order to see how the triangles would be seized by worms, some in a damp state were given to worms kept in confinement. They were seized in three different manners in the case of both the narrow and broad triangles: viz., by the margin; by one of the three angles, which was often completely engulfed in their mouths; and lastly, by suction applied to any part of the flat surface. If lines parallel to the base and an inch apart, are drawn across a triangle with the sides three inches in length, it will be divided into three parts of equal length. Now if worms seized indifferently by chance any part, they would assuredly seize on the basal part or division far oftener than on either of the two other divisions. For the area of the basal to the apical part is as 5 to 1, so that the chance of the former being drawn into a burrow by suction, will be as 5 to 1, compared with the apical part. The base offers two angles and the apex only one, so that the former would have twice as good a chance (independently of the size of the angles) of being engulfed in a worm's mouth, as would the apex. It should, however, be stated that the apical angle is not often seized by worms; the margin at a little distance on either side being preferred. I judge of this from having found in 40 out of 46 cases in which triangles had been drawn into burrows by their apical ends, that the tip had been doubled back within the burrow for a length of between $\frac{1}{20}$th of an inch and 1 inch. Lastly, the proportion between the margins of the basal and apical parts is as 3 to 2 for the broad, and $2\frac{1}{2}$ to 2 for the narrow triangles. From these several considerations it might certainly have been expected, supposing that worms seized hold of the triangles by chance, that a considerably larger proportion would have been dragged into the burrows by the basal than by the apical part; but we shall immediately see how different was the result.

Triangles of the above specified sizes were scattered on the ground in many places and on many successive nights near worm-burrows, from which the leaves, petioles, twigs, &c., with which they had been plugged, were removed. Altogether 303 triangles were drawn by worms into their burrows: 12 others were drawn in by both ends, but as it was impossible to judge by which end they had been first seized, these are excluded. Of the 303, 62 per cent. had been drawn in by the apex (using this term for all drawn in by the apical part, one inch in length); 15 per cent. by the middle; and 23 per cent. by the basal part. If they had been drawn indifferently by any point, the proportion for the apical, middle and basal parts would have been 33·3. per cent. for each; but, as we have just seen, it might have been expected that a much larger proportion would have been drawn in by the basal than by any other part. As the case stands, nearly three times as many were drawn in by the apex as by the base. If we consider the broad triangles by themselves, 59 per cent. were drawn in by the apex, 25 per cent. by the middle, and 16 per cent. by the base. Of the narrow triangles, 65 per cent. were drawn in by the apex, 14 per cent. by the middle, and 21 per cent. by the base; so that here those drawn in by the apex were more than 3 times as many as those drawn in by the base. We may therefore conclude that the manner in which the triangles are drawn into the burrows is not a matter of chance.

In eight cases, two triangles had been drawn into the same burrow, and in seven of these cases, one had been drawn in by the apex and the other by the base. This again indicates that the result is not determined by chance. Worms appear sometimes to revolve in the act of drawing in the triangles, for five out of the whole lot had been wound into an irregular spire round the inside of the burrow. Worms kept in a warm room drew 63 triangles into their burrows; but, as in the case of the pine-leaves, they worked in a rather careless manner, for only 44 per cent. were drawn in by the apex, 22 per cent. by the middle, and

33 per cent. by the base. In five cases, two triangles were drawn into the same burrow.

It may be suggested with much apparent probability that so large a proportion of the triangles were drawn in by the apex, not from the worms having selected this end as the most convenient for the purpose, but from having first tried in other ways and failed. This notion was countenanced by the manner in which worms in confinement were seen to drag about and drop the triangles; but then they were working carelessly. I did not at first perceive the importance of this subject, but merely noticed that the bases of those triangles which had been drawn in by the apex, were generally clean and not crumpled. The subject was afterwards attended to carefully. In the first place several triangles which had been drawn in by the basal angles, or by the base, or a little above the base, and which were thus much crumpled and dirtied, were left for some hours in water and were then well shaken while immersed; but neither the dirt nor the creases were thus removed. Only slight creases could be obliterated, even by pulling the wet triangles several times through my fingers. Owing to the slime from the worms' bodies, the dirt was not easily washed off. We may therefore conclude that if a triangle, before being dragged in by the apex, had been dragged into a burrow by its base with even a slight degree of force, the basal part would long retain its creases and remain dirty. The condition of 89 triangles (65 narrow and 24 broad ones), which had been drawn in by the apex, was observed; and the bases of only 7 of them were at all creased, being at the same time generally dirty. Of the 82 uncreased triangles, 14 were dirty at the base; but it does not follow from this fact that these had first been dragged towards the burrows by their bases; for the worms sometimes covered large portions of the triangles with slime, and these when dragged by the apex over the ground would be dirtied; and during rainy weather, the triangles were often dirtied over one whole side or over both sides. If the worms had dragged the triangles to the mouths of their burrows by their bases, as often as by their apices, and had then perceived, without actually trying to draw them into the burrow, that the broader end was not well adapted for this purpose—even in this case a large proportion would probably have had their basal ends dirtied. We may therefore infer—improbable as is the inference—that worms are able by some means to judge which is the best end by which to draw triangles of paper into their burrows.

The per centage results of the foregoing observations on the manner in which worms draw various kinds of objects into the mouths of their burrows may be abridged as follows :—

Nature of Object.	Drawn into the burrows, by or near the apex.	Drawn in, by or near the middle.	Drawn in, by or near the base.
Leaves of various kinds . . .	80	11	9
——— of the Lime, basal margin of blade broad, apex acuminated	79	17	4
——— of a Laburnum, basal part of blade as narrow as, or sometimes little narrower than the apical part. . .	63	10	27
——— of the Rhododendron, basal part of blade often narrower than the apical part . .	34	..	66
——— of Pine-trees, consisting of two needles arising from a common base . . .	..	..	100

Nature of Object.	Drawn into the burrows, by or near the apex.	Drawn in, by or near the middle.	Drawn in, by or near the base.
Petioles of a Clematis, somewhat pointed at the apex, and blunt at the base . .	76	..	24
——— of the Ash, the thick basal end often drawn in to serve as food	48·5	..	51·5
——— of Robinia, extremely thin, especially towards the apex, so as to be ill-fitted for plugging up the burrows .	44	..	56
Triangles of paper, of the two sizes .	62	15	23
——— of the broad ones alone .	59	25	16
——— of the narrow ones alone .	65	14	21

If we consider these several cases, we can hardly escape from the conclusion that worms show some degree of intelligence in their manner of plugging up their burrows. Each particular object is seized in too uniform a manner, and from causes which we can generally understand, for the result to be attributed to mere chance. That every object has not been drawn in by its pointed end, may be accounted for by labour having been saved through some being inserted by their broader or thicker ends. No doubt worms are led by instinct to plug up their burrows; and it might have been expected that they would have been led by instinct how best to act in each particular case, independently of intelligence. We see how difficult it is to judge whether intelligence comes into play, for even plants might sometimes be thought to be thus directed; for instance when displaced leaves re-direct their upper surfaces towards the light by extremely complicated movements and by the shortest course. With animals, actions appearing due to intelligence may be performed through inherited habit without any intelligence, although aboriginally thus acquired. Or the habit may have been acquired through the preservation and inheritance of beneficial variations of some other habit; and in this case the new habit will have been acquired independently of intelligence throughout the whole course of its development. There is no *à priori* improbability in worms having acquired special instincts through either of these two latter means. Nevertheless it is incredible that instincts should have been developed in reference to objects, such as the leaves or petioles of foreign plants, wholly unknown to the progenitors of the worms which act in the described manner. Nor are their actions so unvarying or inevitable as are most true instincts.

As worms are not guided by special instincts in each particular case, though possessing a general instinct to plug up their burrows, and as chance is excluded, the next most probable conclusion seems to be that they try in many different ways to draw in objects, and at last succeed in some one way. But it is surprising that an animal so low in the scale as a worm should have the capacity for acting in this manner, as many higher animals have no such capacity. For instance, ants may be seen vainly trying to drag an object transversely to their course, which could be easily drawn longitudinally; though after a time they generally act in a wiser manner. M. Fabre states* that a Sphex—an insect belonging to the same highly-endowed order with ants—stocks its nest with paralysed

* See his interesting work, 'Souvenirs entomologiques,' 1879, p. 168–177.

grasshoppers, which are invariably dragged into the burrow by their antennæ. When these were cut off close to the head, the Sphex seized the palpi; but when these were likewise cut off, the attempt to drag its prey into the burrow was given up in despair. The Sphex had not intelligence enough to seize one of the six legs or the ovipositor of the grasshopper, which, as M. Fabre remarks, would have served equally well. So again, if the paralysed prey with an egg attached to it be taken out of the cell, the Sphex after entering and finding the cell empty, nevertheless closes it up in the usual elaborate manner. Bees will try to escape and go on buzzing for hours on a window, one half of which has been left open. Even a pike continued during three months to dash and bruise itself against the glass sides of an aquarium, in the vain attempt to seize minnows on the opposite side.* A cobra-snake was seen by Mr. Layard † to act much more wisely than either the pike or the Sphex;

* Möbius, 'Die Bewegungen der Thiere,' &c., 1873, p. 111.

† 'Annals and Mag. of N. History,' series ii. vol. ix. 1852, p. 333.

it had swallowed a toad lying within a hole, and could not withdraw its head; the toad was disgorged, and began to crawl away; it was again swallowed and again disgorged; and now the snake had learnt by experience, for it seized the toad by one of its legs and drew it out of the hole. The instincts of even the higher animals are often followed in a senseless or purposeless manner: the weaver-bird will perseveringly wind threads through the bars of its cage, as if building a nest: a squirrel will pat nuts on a wooden floor, as if he had just buried them in the ground: a beaver will cut up logs of wood and drag them about, though there is no water to dam up; and so in many other cases.

Mr. Romanes who has specially studied the minds of animals, believes that we can safely infer intelligence, only when we see an individual profiting by its own experience. By this test the cobra showed some intelligence; but this would have been much plainer if on a second occasion he had drawn a toad out of a hole by its leg. The Sphex failed signally in this respect. Now if worms try to drag objects into their burrows first in one way and then in another, until they at last succeed, they profit, at least in each particular instance, by experience.

But evidence has been advanced showing that worms do not habitually try to draw objects into their burrows in many different ways. Thus half-decayed lime-leaves from their flexibility could have been drawn in by their middle or basal parts, and were thus drawn into the burrows in considerable numbers; yet a large majority were drawn in by or near the apex. The petioles of the Clematis could certainly have been drawn in with equal ease by the base and apex; yet three times and in certain cases five times as many were drawn in by the apex as by the base. It might have been thought that the foot-stalks of leaves would have tempted the worms as a convenient handle; yet they are not largely used, except when the base of the blade is narrower than the apex. A large number of the petioles of the ash are drawn in by the base; but this part serves the worms as food. In the case of pine-leaves worms plainly show that they at least do not seize the leaf by chance; but their choice does not appear to be determined by the divergence of the two needles, and the consequent advantage or necessity of drawing them into their burrows by the base. With respect to the triangles of paper, those which had been drawn in by the apex rarely had their bases creased or dirty; and this shows that the worms had not often first tried to drag them in by this end.

If worms are able to judge, either before drawing or after having drawn an object close to the mouths of their burrows, how best to drag it in, they must acquire some notion of its general shape. This they probably acquire by touching it in many places with the anterior extremity of their bodies, which serves as a tactile organ. It may be well to remember how perfect the sense of touch becomes in a man when born blind and deaf, as are worms. If worms have the power of acquiring some notion, however rude, of the shape of an object and of their burrows, as seems to be the case, they deserve to be called intelligent; for they then act in nearly the same manner as would a man under similar circumstances.

To sum up, as chance does not determine the manner in which objects are drawn into the burrows, and as the existence of specialized instincts for each particular case cannot be admitted, the first and most natural supposition is that worms try all methods until they at last succeed; but many appearances are opposed to such a supposition. One alternative alone is left, namely, that worms, although standing low in the scale of organization, possess some degree of intelligence. This will strike every one as very improbable; but it may be doubted whether we know enough about the nervous system of the lower animals to justify our natural distrust of such a conclusion. With respect to the small size of the cerebral ganglia, we should remember what a mass of inherited knowledge, with some power of adapting means to an end, is crowded into the minute brain of a worker-ant.

Means by which worms excavate their burrows.—This is effected in two ways; by pushing away the earth on all sides, and by swallowing it. In the former case, the worm inserts the stretched out and attenuated anterior extremity of its body into any little crevice, or hole; and then, as Perrier remarks,* the pharynx is pushed forwards into this part, which consequently swells and pushes away the earth on all sides. The anterior extremity thus serves as a wedge. It also serves, as we have before seen, for prehension and suction, and as a tactile organ. A worm was placed on loose mould, and it buried itself in between two and three minutes. On another occasion four worms disappeared in 15 minutes between the sides of the pot and the earth, which had been moderately pressed down. On a third occasion three large worms and a small one were placed on loose mould well mixed with fine sand and firmly pressed down, and they all disappeared, except the tail of one, in 35 minutes. On a fourth occasion six large worms were placed on argillaceous mud mixed with sand firmly pressed down, and they disappeared, except the extreme tips of the tails of two of them, in 40 minutes. In none of these cases, did the worms swallow, as far as could be seen, any earth. They

* 'Archives de Zoolog. expér.' tom. iii. 1874, p. 405.

generally entered the ground close to the sides of the pot.

A pot was next filled with very fine ferruginous sand, which was pressed down, well watered, and thus rendered extremely compact. A large worm left on the surface did not succeed in penetrating it for some hours, and did not bury itself completely until 25 hrs. 40 min. had elapsed. This was effected by the sand being swallowed, as was evident by the large quantity ejected from the vent, long before the whole body had disappeared. Castings of a similar nature continued to be ejected from the burrow during the whole of the following day.

As doubts have been expressed by some writers whether worms ever swallow earth solely for the sake of making their burrows, some additional cases may be given. A mass of fine reddish sand, 23 inches in thickness, left on the ground for nearly two years, had been penetrated in many places by worms; and their castings consisted partly of the reddish sand and partly of black earth brought up from beneath the mass. This sand had been dug up from a considerable depth, and was of so poor a nature that weeds could not grow on it. It is therefore highly improbable that it should have been swallowed by the worms as food. Again in a field near my house the castings frequently consist of almost pure chalk, which lies at only a little depth beneath the surface; and here again it is very improbable that the chalk should have been swallowed for the sake of the very little organic matter which could have percolated into it from the poor overlying pasture. Lastly, a casting thrown up through the concrete and decayed mortar between the tiles, with which the now ruined aisle of Beaulieu Abbey had formerly been paved, was washed, so that the coarser matter alone was left. This consisted of grains of quartz, micaceous slate, other rocks, and bricks or tiles, many of them from $\frac{1}{20}$ to $\frac{1}{10}$ inch in diameter. No one will suppose that these grains were swallowed as food, yet they formed more than half of the casting, for they weighed 19 grains, the whole casting having weighed 33 grains. Whenever a worm burrows to a depth of some feet in undisturbed compact ground, it must form its passage by swallowing the earth; for it is incredible that the ground could yield on all sides to the pressure of the pharynx when pushed forwards within the worm's body.

That worms swallow a larger quantity of earth for the sake of extracting any nutritious matter which it may contain than for making their burrows, appears to me certain. But as this old belief has been doubted by so high an authority as Claparède, evidence in its favour must be given in some detail. There is no *à priori* improbability in such a belief, for besides other annelids, especially the *Arenicola marina*, which throws up such a profusion of castings on our tidal sands, and which it is believed thus subsists, there are animals belonging to the most distinct classes, which do not burrow, but habitually swallow large quantities of sand; namely the molluscan Onchidium and many Echinoderms.*

If earth were swallowed only when worms deepened their burrows or made new ones, castings would be thrown up only occasionally; but in many places fresh castings may

* I state this on the authority of Semper, 'Reisen im Archipel der Philippinen," Th. ii. 1877, p. 30.

be seen every morning, and the amount of earth ejected from the same burrow on successive days is large. Yet worms do not burrow to a great depth, except when the weather is very dry or intensely cold. On my lawn the black vegetable mould is only about 5 inches in thickness, and overlies light-coloured or reddish clayey soil: now when castings are thrown up in the greatest profusion, only a small proportion are light coloured, and it is incredible that the worms should daily make fresh burrows in every direction in the thin superficial layer of dark-coloured humus, unless they obtained nutriment of some kind from it. I have observed a strictly analogous case in a field near my house where bright red clay lay close beneath the surface. Again on one part of the Downs near Winchester the vegetable mould overlying the chalk was found to be only from 3 to 4 inches in thickness; and the many castings here ejected were as black as ink and did not effervesce with acids; so that the worms must have confined themselves to this thin superficial layer of mould, of which large quantities were daily swallowed. In another place at no great distance the castings were white; and why the worms should have burrowed into the chalk in some places and not in others, I am unable to conjecture.

Two great piles of leaves had been left to decay in my grounds, and months after their removal, the bare surface, several yards in diameter, was so thickly covered during several months with castings that they formed an almost continuous layer; and the large number of worms which lived here must have subsisted during these months on nutritious matter contained in the black earth.

The lowest layer from another pile of decayed leaves mixed with some earth was examined under a high power, and the number of spores of various shapes and sizes which it contained was astonishingly great; and these crushed in the gizzards of worms may largely aid in supporting them. Whenever castings are thrown up in the greatest number, few or no leaves are drawn into the burrows; for instance the turf along a hedgerow, about 200 yards in length, was daily observed in the autumn during several weeks, and every morning many fresh castings were seen; but not a single leaf was drawn into these burrows. These castings from their blackness and from the nature of the subsoil could not have been brought up from a greater depth than 6 or 8 inches. On what could these worms have subsisted during this whole time, if not on matter contained in the black earth? On the other hand, whenever a large number of leaves are drawn into the burrows, the worms seem to subsist chiefly on them, for few earth-castings are then ejected on the surface. This difference in the behaviour of worms at different times, perhaps explains a statement by Claparède, namely, that triturated leaves and earth are always found in distinct parts of their intestines.

Worms sometimes abound in places where they can rarely or never obtain dead or living leaves; for instance, beneath the pavement in well-swept courtyards, into which leaves are only occasionally blown. My son Horace examined a house, one corner of which had subsided; and he found here in the cellar, which was extremely damp, many small worm-castings thrown up between the stones with which the cellar was paved; and in this case it is improbable that the worms could ever have obtained leaves.

But the best evidence, known to me, of worms subsisting for at least considerable periods of time solely on the organic matter contained in earth, is afforded by some facts communicated to me by Dr. King. Near Nice large castings abound in extraordinary numbers, so that 5 or 6 were often found within the space of a square foot. They consist of fine, pale-coloured earth, containing calcareous matter, which after having passed through the bodies of worms and being dried, coheres with considerable force. I have reason to believe that these castings had been formed by species of Perichæta, which have been naturalised here from the East.* They

* Dr. King gave me some worms collected near Nice, which, as he believes, had constructed these castings. They were sent to M. Perrier, who with great kindness examined and named them for me: they consisted of *Perichæta affinis*, a native of Cochin China and of the Philippines; *P. Luzonica*, a native of Luzon in the Philippines; and *P. Houlleti*, which lives near Calcutta. M. Perrier informs me that species of Perichæta have been naturalized in the gardens near Montpellier and in Algiers. Before I had any reason to suspect that the tower-like castings from Nice had been formed by worms not endemic in the country, I was

rise like towers (see Fig. 2), with their summits often a little broader than their bases,

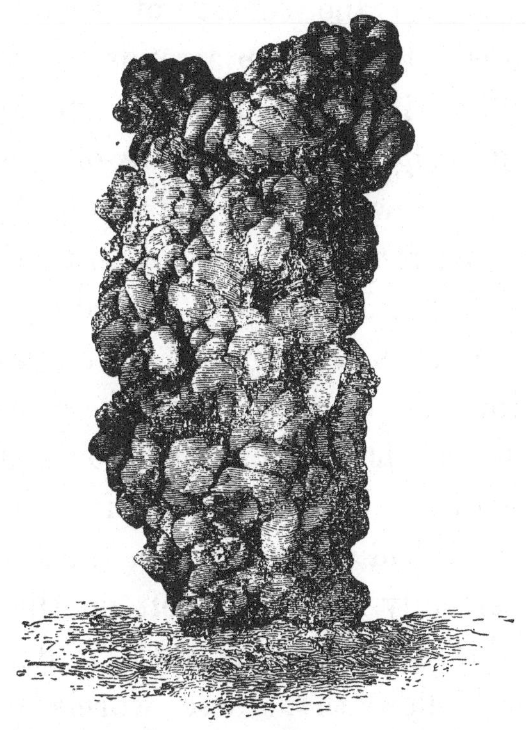

Fig. 2.

Tower-like casting from near Nice, constructed of earth, voided probably by a species of Perichæta: of natural size, copied from a photograph.

sometimes to a height of above 3 and often to a height of $2\frac{1}{2}$ inches. The tallest of those

---

greatly surprised to see how closely they resembled castings sent to me from near Calcutta, where it is known that species of Perichæta abound.

which were measured was 3·3 inch in height and 1 in diameter. A small cylindrical passage runs up the centre of each tower, through which the worm ascends to eject the earth which it has swallowed, and thus to add to its height. A structure of this kind would not allow leaves being easily dragged from the surrounding ground into the burrows; and Dr. King, who looked carefully, never saw even a fragment of a leaf thus drawn in. Nor could any trace be discovered of the worms having crawled down the exterior surfaces of the towers in search of leaves; and had they done so, tracks would almost certainly have been left on the upper part whilst it remained soft. It does not, however, follow that these worms do not draw leaves into their burrows during some other season of the year, at which time they would not build up their towers.

From the several foregoing cases, it can hardly be doubted that worms swallow earth, not only for the sake of making their burrows, but for obtaining food. Hensen, however, concludes from his analyses of humus that worms probably could not live on ordinary vegetable mould, though he admits that they might be nourished to some extent by leaf-mould.* But we have seen that worms eagerly devour raw meat, fat, and dead worms; and ordinary mould can hardly fail to contain many ova, larvæ, and small living or dead creatures, spores of cryptogamic plants, and micrococci, such as those which give rise to saltpetre. These various organisms, together with some cellulose from any leaves and roots not utterly decayed, might well account for such large quantities of mould being swallowed by worms. It may be worth while here to recall the fact that certain species of Utricularia, which grow in damp places in the tropics, possess bladders beautifully constructed for catching minute subterranean animals; and these traps would not have been developed unless many small animals inhabited such soil.

The depth to which worms penetrate, and the construction of their burrows.— Although worms usually live near the surface, yet they burrow to a considerable depth during long-

* 'Zeitschrift für wissenschaft. Zoolog.' B. xxviii. 1877, p. 364.

continued dry weather and severe cold. In Scandinavia, according to Eisen, and in Scotland, according to Mr. Lindsay Carnagie, the burrows run down to a depth of from 7 to 8 feet; in North Germany, according to Hoffmeister, from 6 to 8 feet, but Hensen says, from 3 to 6 feet. This latter observer has seen worms frozen at a depth of 1½ feet beneath the surface. I have not myself had many opportunities for observation, but I have often met with worms at depths of 3 to 4 feet. In a bed of fine sand overlying the chalk, which had never been disturbed, a worm was cut into two at 55 inches, and another was found here in December at the bottom of its burrow, at 61 inches beneath the surface. Lastly, in earth near an old Roman Villa, which had not been disturbed for many centuries, a worm was met with at a depth of 66 inches; and this was in the middle of August.

The burrows run down perpendicularly, or more commonly a little obliquely. They are said sometimes to branch, but as far as I have seen this does not occur, except in recently dug ground and near the surface. They are generally, or as I believe invariably, lined with a thin layer of fine, dark-coloured earth voided by the worms; so that they must at first be made a little wider than their ultimate diameter. I have seen several burrows in undisturbed sand thus lined at a depth of 4 ft. 6 in.; and others close to the surface thus lined in recently dug ground. The walls of fresh burrows are often dotted with little globular pellets of voided earth, still soft and viscid; and these, as it appears, are spread out on all sides by the worm as it travels up or down its burrow. The lining thus formed becomes very compact and smooth when nearly dry, and closely fits the worm's body. The minute reflexed bristles which project in rows on all sides from the body, thus have excellent points of support; and the burrow is rendered well adapted for the rapid movement of the animal. The lining appears also to strengthen the walls, and perhaps saves the worm's body from being scratched. I think so because several burrows which passed through a layer of sifted coal-cinders, spread over turf to a thickness of $1\frac{1}{2}$ inch, had been thus lined to an unusual thickness. In this case the worms, judging from the castings, had pushed the cinders away on all sides and had not swallowed any of them. In another place, burrows similarly lined, passed through a layer of coarse coal-cinders, 3½ inches in thickness. We thus see that the burrows are not mere excavations, but may rather be compared with tunnels lined with cement.

The mouths of the burrow are in addition often lined with leaves; and this is an instinct distinct from that of plugging them up, and does not appear to have been hitherto noticed. Many leaves of the Scotch-fir or pine (*Pinus sylvestris*) were given to worms kept in confinement in two pots; and when after several weeks the earth was carefully broken up, the upper parts of three oblique burrows were found surrouuded for lengths of 7, 4, and 3½ inches with pine-leaves, together with fragments of other leaves which had been given the worms as food. Glass beads and bits of tile, which had been strewed on the surface of the soil, were stuck into the interstices between the pine-leaves; and these interstices were likewise plastered with the viscid castings voided by the worms. The structures thus formed cohered so well, that I succeeded in removing one with only a little earth adhering to it. It consisted of a slightly curved cylindrical case, the interior of which could be seen through holes in the sides and at either end. The pine-leaves had all been drawn in by their bases; and the sharp points of the needles had been pressed into the lining of voided earth. Had this not been effectually done, the sharp points would have prevented the retreat of the worms into their burrows; and these structures would have resembled traps armed with converging points of wire, rendering the ingress of an animal easy and its egress difficult or impossible. The skill shown by these worms is noteworthy and is the more remarkable, as the Scotch pine is not a native of this district.

After having examined these burrows made by worms in confinement, I looked at those in a flower-bed near some Scotch pines. These had all been plugged up in the ordinary manner with the leaves of this tree, drawn in for a length of from 1 to 1½ inch; but the mouths of many of them were likewise lined with them, mingled with fragments of other kinds of leaves, drawn in to a depth of 4 or 5 inches. Worms often remain, as formerly stated, for a long time close to the mouths of their burrows, apparently for warmth; and the basket-like structures formed of leaves would keep their bodies from coming into close contact with the cold damp earth. That they habitually rested on the pine-leaves, was rendered probable by their clean and almost polished surfaces.

The burrows which run far down into the ground, generally, or at least often, terminate in a little enlargement or chamber. Here, according to Hoffmeister, one or several worms pass the winter rolled up into a ball. Mr. Lindsay Carnagie informed me (1838) that he had examined many burrows over a stone-quarry in Scotland, where the overlying boulder-clay and mould had recently been cleared away, and a little vertical cliff thus left. In several cases the same burrow was a little enlarged at two or three points one beneath the other; and all the burrows terminated in a rather large chamber, at a depth of 7 or 8 feet from the surface. These chambers contained many small sharp bits of stone and husks of flax-seeds. They must also have contained living seeds, for on the following spring Mr. Carnagie saw grass-plants sprouting out of some of the intersected chambers. I found at Abinger in Surrey two burrows terminating in similar chambers at a depth of 36 and 41 inches, and these were lined or paved with little pebbles, about as large as mustard seeds; and in one of the chambers there was a decayed oat-grain, with its husk. Hensen likewise states that the bottoms of the burrows are lined with little stones; and where these could not be procured, seeds, apparently of the pear, had been used, as many as fifteen having been carried down into a single burrow, one of which had germinated.* We thus see how easily a botanist might be deceived who wished to learn how long deeply buried seeds remained alive, if he were to collect earth from a considerable depth, on the supposition that it could contain only seeds which had long lain buried. It is probable that the little stones,

* 'Zeitschrift für wissenschaft. Zoolog.' B. xxviii. 1877, p. 356.

as well as the seeds, are carried down from the surface by being swallowed; for a surprising number of glass beads, bits of tile and of glass were certainly thus carried down by worms kept in pots; but some may have been carried down within their mouths. The sole conjecture which I can form why worms line their winter-quarters with little stones and seeds, is to prevent their closely coiled-up bodies from coming into close contact with the surrounding cold soil; and such contact would perhaps interfere with their respiration which is effected by the skin alone.

A worm after swallowing earth, whether for making its burrow or for food, soon comes to the surface to empty its body. The ejected earth is thoroughly mingled with the intestinal secretions, and is thus rendered viscid. After being dried it sets hard. I have watched worms during the act of ejection, and when the earth was in a very liquid state it was ejected in little spurts, and when not so liquid by a slow peristaltic movement. It is not cast indifferently on any side, but with some care, first on one and then on another side; the tail being used almost like a trowel.

As soon as a little heap is formed, the worm apparently avoids, for the sake of safety, protruding its tail; and the earthy matter is forced up through the previously deposited soft mass. The mouth of the same burrow is used for this purpose for a considerable time. In the case of the tower-like castings (see Fig. 2) near Nice, and of the similar but still taller towers from Bengal (hereafter to be described and figured) a considerable degree of skill is exhibited in their construction. Dr. King also observed that the passage up these towers hardly ever ran in the same exact line with the underlying burrow, so that a thin cylindrical object such as a haulm of grass, could not be passed down the tower into the burrow; and this change of direction probably serves in some manner as a protection. When a worm comes to the surface to eject earth, the tail protrudes, but when it collects leaves its head must protrude. Worms therefore must have the power of turning round in their closely-fitting burrows; and this, as it appears to us, would be a difficult feat.

Worms do not always eject their castings on

the surface of the ground. When they can find any cavity, as when burrowing in newly turned-up earth, or between the stems of banked-up plants, they deposit their castings in such places. So again any hollow beneath a large stone lying on the surface of the ground, is soon filled up with their castings. According to Hensen, old burrows are habitually used for this purpose; but as far as my experience serves, this is not the case, excepting with those near the surface in recently dug ground. I think that Hensen may have been deceived by the walls of old burrows, lined with black earth, having sunk in or collapsed; for black streaks are thus left, and these are conspicuous when passing through light-coloured soil, and might be mistaken for completely filled-up burrows.

It is certain that old burrows collapse in the course of time; for as we shall see in the next chapter, the fine earth voided by worms, if spread out uniformly, would form in many places in the course of a year a layer $\frac{1}{5}$ of an inch in thickness; so that at any rate this large amount is not deposited within the old unused burrows. If the burrows did not collapse, the whole ground would be first thickly riddled with holes to a depth of about ten inches, and in fifty years a hollow unsupported space, ten inches in depth, would be left. The holes left by the decay of successively formed roots of trees and plants must likewise collapse in the course of time.

The burrows of worms run down perpendicularly or a little obliquely, and where the soil is at all argillaceous, there is no difficulty in believing that the walls would slowly flow or slide inwards during very wet weather. When, however, the soil is sandy or mingled with many small stones, it can hardly be viscous enough to flow inwards during even the wettest weather; but another agency may here come into play. After much rain the ground swells, and as it cannot expand laterally, the surface rises; during dry weather it sinks again. For instance, a large flat stone laid on the surface of a field sank 3·33 mm. whilst the weather was dry between May 9th and June 13th, and rose 1·91 mm. between September 7th and 19th, much rain having fallen during the latter part of this time. During frosts and thaws the movements were twice as great. These observations were made by my son Horace, who will hereafter publish an account of the movements of this stone during successive wet and dry seasons, and of the effects of its being undermined by worms. Now when the ground swells, if it be penetrated by cylindrical holes, such as worm-burrows, their walls will tend to yield and be pressed inwards; and the yielding will be greater in the deeper parts (supposing the whole to be equally moistened) from the greater weight of the superincumbent soil which has to be raised, than in the parts near the surface. When the ground dries, the walls will shrink a little and the burrows will be a little enlarged. Their enlargement, however, through the lateral contraction of the ground, will not be favoured, but rather opposed, by the weight of the superincumbent soil.

Distribution of Worms.—Earth-worms are found in all parts of the world, and some of the genera have an enormous range.* They inhabit the most isolated islands; they

* Perrier, 'Archives de Zoolog. expér.' tom. 3, p. 378, 1874.

abound in Iceland, and are known to exist in the West Indies, St. Helena, Madagascar, New Caledonia and Tahiti. In the Antarctic regions, worms from Kerguelen Land have been described by Ray Lankester; and I found them in the Falkland Islands. How they reach such isolated islands is at present quite unknown. They are easily killed by salt-water, and it does not appear probable that young worms or their egg-capsules could be carried in earth adhering to the feet or beaks of land-birds. Moreover Kerguelen Land is not now inhabited by any land-bird.

In this volume we are chiefly concerned with the earth cast up by worms, and I have gleaned a few facts on this subject with respect to distant lands. Worms throw up plenty of castings in the United States. In Venezuela, castings, probably ejected by species of Urochæta, are common in the gardens and fields, but not in the forests, as I hear from Dr. Ernst of Caracas. He collected 156 castings from the court-yard of his house, having an area of 200 square yards. They varied in bulk from half a cubic centimeter to five cubic centimeters, and were on an average three cubic centimeters. They were, therefore of small size in comparison with those often found in England; for six large castings from a field near my house averaged 16 cubic centimeters. Several species of earth-worms are common in St. Catharina in South Brazil, and Fritz Müller informs me "that in most parts of "the forests and pasture-lands, the whole soil, "to a depth of a quarter of a metre, looks as if it "had passed repeatedly through the intestines "of earth-worms, even where hardly any castings are to be seen on the surface." A gigantic but very rare species is found there, the burrows of which are sometimes even two centimeters or nearly $\frac{4}{5}$ of an inch in diameter, and which apparently penetrate the ground to a great depth.

In the dry climate of New South Wales, I hardly expected that worms would be common; but Dr. G. Krefft of Sydney, to whom I applied, after making enquiries from gardeners and others, and from his own observations, informs me that their castings abound. He sent me some collected after heavy rain, and they consisted of little pellets, about ·15 inch in diameter; and the blackened sandy earth of which they were formed still cohered with considerable tenacity.

The late Mr. John Scott of the Botanic Gardens near Calcutta made many observations for me on worms living under the hot and humid climate of Bengal. The castings abound almost everywhere, in jungles and in the open ground, to a greater degree, as he thinks, than in England. After the water has subsided from the flooded rice-fields, the whole surface very soon becomes studded with castings—a fact which much surprised Mr. Scott, as he did not know how long worms could survive beneath water. They cause much trouble in the Botanic garden, "for "some of the finest of our lawns can be kept "in anything like order only by being almost "daily rolled; if left undisturbed for a few days "they become studded with large castings." These closely resemble those described as abounding near Nice; and they are probably the work of a species of Perichæta. They stand up like towers, with an open passage in the centre.

A figure of one of these castings from a photograph is here given (Fig. 3). The largest received by me was $3\frac{1}{2}$ inches in height and 1·35 inch in diameter; another

Fig. 3.

A tower-like casting, probably ejected by a species of Perichæta, from the Botanic Garden, Calcutta: of natural size, engraved from a photograph.

was only $\frac{3}{4}$ inch in diameter and $2\frac{3}{4}$ in height.

In the following year, Mr. Scott measured several of the largest; one was 6 inches in height and nearly $1\frac{1}{2}$ in diameter: two others were 5 inches in height and respectively 2 and rather more than $2\frac{1}{2}$ inches in diameter. The average weight of the 22 castings sent to me was 35 grammes ($1\frac{1}{4}$ oz.); and one of them weighed 44·8 grammes (or 2 oz.). All these castings were thrown up either in one night or in two. Where the ground in Bengal is dry, as under large trees, castings of a different kind are found in vast numbers: these consist of little oval or conical bodies, from about the $\frac{1}{20}$ to rather above $\frac{1}{10}$ of an inch in length. They are obviously voided by a distinct species of worms.

The period during which worms near Calcutta display such extraordinary activity lasts for only a little over two months, namely, during the cool season after the rains. At this time they are generally found within about 10 inches beneath the surface. During the hot season they burrow to a greater depth, and are then found coiled up and apparently hybernating. Mr. Scott has never seen them at a greater depth than $2\frac{1}{2}$ feet, but has heard of their having been found at 4 feet. Within the forests, fresh castings may be found even during the hot season. The worms in the Botanic garden, during the cool and dry season, draw many leaves and little sticks into the mouths of their burrows, like our English worms; but they rarely act in this manner during the rainy season.

Mr. Scott saw worm-castings on the lofty mountains of Sikkim in North India. In South India Dr. King found in one place, on the plateau of the Nilgiris, at an elevation of 7000 feet, "a good many castings," which are interesting for their great size. The worms which eject them are seen only during the wet season, and are reported to be from 12 to 15 inches in length, and as thick as a man's little finger. These castings were collected by Dr. King after a period of 110 days without any rain; and they must have been ejected either during the north-east or more probably during the previous south-west monsoon; for their surfaces had suffered some disintegration and they were penetrated by many fine roots. A drawing is here given (Fig. 4) of one which seems to have best retained its original size and appearance. Notwithstanding some loss from disintegration, five of the largest of these castings (after having been well sun-dried) weighed each on an average 89·5 grammes,

Fig. 4.

A casting from the Nilgiri Mountains in South India; of natural size, engraved from a photograph.

or above 3 oz.; and the largest weighed 123·14 grammes, or $4\frac{1}{3}$ oz.,—that is above a quarter of a pound! The largest convolutions were rather more than one inch in diameter; but it is probable that they had subsided a little whilst soft, and that their diameters had thus been increased. Some had flowed so much that they now consisted of a pile of almost flat confluent cakes. All were formed of fine, rather light-coloured earth, and were surprisingly hard and compact, owing no doubt to the animal matter by which the particles of earth had been cemented together. They did not disintegrate, even when left for some hours in water. Although they had been cast up on the surface of gravelly soil, they contained extremely few bits of rock, the largest of which was only ·15 inch in diameter.

Dr. King saw in Ceylon a worm about 2 feet in length and $\frac{1}{2}$ inch in diameter; and he was told that it was a very common species during the wet season. These worms must throw up castings at least as large as those on the Nilgiri Mountains; but Dr. King saw none during his short visit to Ceylon. Sufficient facts have now been given, showing that worms do much work in bringing up fine earth to the surface in most or all parts of the world, and under the most different climates.

CHAPTER III.

THE AMOUNT OF FINE EARTH BROUGHT UP BY WORMS TO THE SURFACE.

Rate at which various objects strewed on the surface of grass-fields are covered up by the castings of worms—The burial of a paved path—The slow subsidence of great stones left on the surface—The number of worms which live within a given space—The weight of earth ejected from a burrow, and from all the burrows within a given space—The thickness of the layer of mould which the castings on a given space would form within a given time if uniformly spread out—The slow rate at which mould can increase to a great thickness—Conclusion.

We now come to the more immediate subject of this volume, namely the amount of earth which is brought up by worms from beneath the surface, and is afterwards spread out more or less completely by the rain and wind. The amount can be judged of by two methods,—by the rate at which objects left on the surface are buried, and more accurately by weighing the quantity brought up within a given time. We will begin with the first method, as it was first followed.

Near Maer Hall in Staffordshire, quick-lime had been spread about the year 1827 thickly over a field of good pasture-land, which had not since been ploughed. Some square holes were dug in this field in the beginning of October 1837; and the sections showed a layer of turf, formed by the matted roots of the grasses, ½ inch in thickness, beneath which, at a depth of 2½ inches (or 3 inches from the surface), a layer of the lime in powder or in small lumps could be distinctly seen running all round the vertical sides of the holes. The soil beneath the layer of lime was either gravelly or of a coarse sandy nature, and differed considerably in appearance from the overlying dark-coloured fine mould. Coal-cinders had been spread over a part of this same field either in the year 1833 or 1834; and when the above holes were dug, that is after an interval of 3 or 4 years, the cinders formed a line of black spots round the holes, at a depth of 1 inch beneath the surface, parallel to and above the white layer of lime. Over another part of this field cinders had been strewed, only about half-a-year before, and these either still lay on the surface or were entangled among the roots of the grasses; and I here saw the commencement of the burying process, for worm-castings had been heaped on several of the smaller fragments. After an interval of $4\frac{3}{4}$ years this field was re-examined, and now the two layers of lime and cinders were found almost everywhere at a greater depth than before by nearly 1 inch, we will say by $\frac{3}{4}$ of an inch. Therefore mould to an average thickness of ·22 of an inch had been annually brought up by the worms, and had been spread over the surface of this field.

Coal-cinders had been strewed over another field, at a date which could not be positively ascertained, so thickly that they formed (October, 1837) a layer, 1 inch in thickness at a depth of about 3 inches from the surface. The layer was so continuous that the overlying dark vegetable mould was connected with the sub-soil of red clay only by the roots of the grasses; and when these were broken, the mould and the red clay fell apart. In a third field, on which coal-cinders and burnt marl had been strewed several times at unknown dates, holes were dug in 1842; and a layer of cinders could be traced at a depth of 3½ inches, beneath which at a depth of 9½ inches from the surface there was a line of cinders together with burnt marl. On the sides of one hole there were two layers of cinders, at 2 and 3½ inches beneath the surface; and below them at a depth in parts of 9½, and in other parts of 10½ inches there were fragments of burnt marl. In a fourth field two layers of lime, one above the other, could be distinctly traced, and beneath them a layer of cinders and burnt marl at a depth of from 10 to 12 inches below the surface.

A piece of waste, swampy land was enclosed, drained, ploughed, harrowed and thickly covered in the year 1822 with burnt marl and cinders. It was sowed with grass seeds, and now supports a tolerably good but coarse pasture. Holes were dug in this field in 1837, or 15 years after its reclamation, and we see in the accompanying diagram (Fig. 5), reduced to half of the natural scale, that the turf was ½ inch thick, beneath which there was a layer of vegetable mould 2½ inches thick. This layer did not contain fragments of any kind; but beneath it there was a layer of mould, 1½ inch in thickness, full of fragments

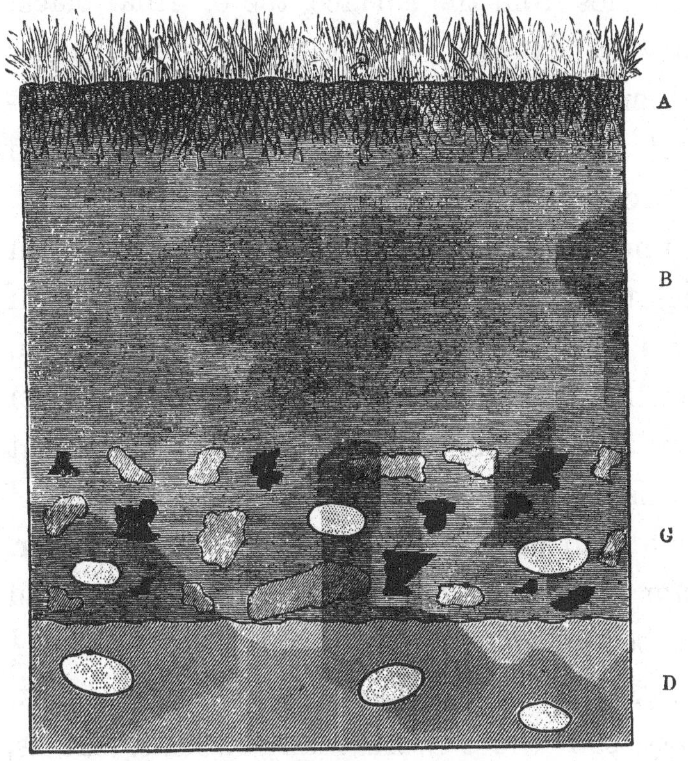

Fig. 5.

Section, reduced to half the natural scale, of the vegetable mould in a field, drained and reclaimed fifteen years previously; A, turf; B, vegetable mould without any stones; C, mould with fragments of burnt marl, coal-cinders and quartz pebbles; D, sub-soil of black, peaty sand with quartz pebbles.

of burnt marl, conspicuous from their red colour, one of which near the bottom was an inch in length; and other fragments of coal-cinders together with a few white quartz pebbles. Beneath this layer and at a depth of 4½ inches from the surface, the original black, peaty, sandy soil with a few quartz pebbles was encountered. Here therefore the fragments of burnt marl and cinders had been covered in the course of 15 years by a layer of fine vegetable mould, only 2½ inches in thickness, excluding the turf. Six and a half years subsequently this field was re-examined, and the fragments were now found at from 4 to 5 inches beneath the surface. So that in this interval of 6½ years, about 1½ inch of mould had been added to the superficial layer. I am surprised that a greater quantity had not been brought up during the whole 21½ years, for in the closely underlying black, peaty soil there were many worms. It is, however, probable that formerly, whilst the land remained poor, worms were scanty; and the mould would then have accumulated slowly. The average annual increase of thickness for the whole period is 1·9 of an inch.

Two other cases are worth recording. In the spring of 1835, a field, which had long existed as poor pasture and was so swampy that it trembled slightly when stamped on, was thickly covered with red sand so that the whole surface appeared at first bright red. When holes were dug in this field after an interval of about 2½ years, the sand formed a layer at a depth of ¾ in. beneath the surface. In 1842 (i.e., 7 years after the sand had been laid on) fresh holes were dug, and now the red sand formed a distinct layer, 2 inches beneath the surface, or 1½ inch beneath the turf; so that on an average, ·21 inches of mould had been annually brought to the surface. Immediately beneath the layer of red sand, the original substratum of black sandy peat extended.

A grass field, likewise not far from Maer Hall, had formerly been thickly covered with marl, and was then left for several years as pasture; it was afterwards ploughed. A friend had three trenches dug in this field 28 years after the application of the marl,*

* This case is given in a postscript to my paper in the 'Transact. Geolog. Soc.' (Vol. v. p. 505), and contains a serious error, as in the account received I mistook the figure 30 for 80. The tenant, moreover, formerly said that he had marled the field thirty years before, but was now positive that this was done in

and a layer of the marl fragments could be traced at a depth, carefully measured, of 12 inches in some parts, and of 14 inches in other parts. This difference in depth depended on the layer being horizontal, whilst the surface consisted of ridges and furrows from the field having been ploughed. The tenant assured me that it had never been turned up to a greater depth than from 6 to 8 inches; and as the fragments formed an unbroken horizontal layer from 12 to 14 inches beneath the surface, these must have been buried by the worms whilst the land was in pasture before it was ploughed, for otherwise they would have been indiscriminately scattered by the plough throughout the whole thickness of the soil. Four-and-a-half years afterwards I had three holes dug in this field, in which potatoes had been lately planted, and the layer of marl-fragments was now found 13 inches beneath the bottoms of the furrows, and therefore probably 15 inches

---

1809, that is twenty-eight years before the first examination of the field by my friend. The error, as far as the figure 80 is concerned, was corrected in an article by me, in the 'Gardeners' Chronicle,' 1844, p. 218.

beneath the general level of the field. It should, however, be observed that the thickness of the blackish sandy soil, which had been thrown up by the worms above the marl-fragments in the course of 32½ years, would have measured less than 15 inches, if the field had always remained as pasture, for the soil would in this case have been much more compact. The fragments of marl almost rested on an undisturbed sub-stratum of white sand with quartz pebbles; and as this would be little attractive to worms, the mould would hereafter be very slowly increased by their action.

We will now give some cases of the action of worms, on land differing widely from the dry sandy or the swampy pastures just described. The chalk formation extends all round my house in Kent; and its surface, from having been exposed during an immense period to the dissolving action of rain-water, is extremely irregular, being abruptly festooned and penetrated by many deep well-like cavities.* During the dissolution of the

* These pits or pipes are still in process of formation. During the last forty years I have seen or heard of five cases, in which a

chalk, the insoluble matter, including a vast number of unrolled flints of all sizes, has

---

circular space, several feet in diameter, suddenly fell in, leaving on the field an open hole with perpendicular sides, some feet in depth. This occurred in one of my own fields, whilst it was being rolled, and the hinder quarters of the shaft horse fell in; two or three cart-loads of rubbish were required to fill up the hole. The subsidence occurred where there was a broad depression, as if the surface had fallen in at several former periods. I heard of a hole which must have been suddenly formed at the bottom of a small shallow pool, where sheep had been washed during many years, and into which a man thus occupied fell to his great terror. The rain-water over this whole district sinks perpendicularly into the ground, but the chalk is more porous in certain places than in others. Thus the drainage from the overlying clay is directed to certain points, where a greater amount of calcareous matter is dissolved than elsewhere. Even narrow open channels are sometimes formed in the solid chalk. As the chalk is slowly dissolved over the whole country, but more in some parts than in others, the undissolved residue—that is the overlying mass of red clay with flints,—likewise sinks slowly down, and tends to fill up the pipes or cavities. But the upper part of the red clay holds together, aided probably by the roots of plants, for a longer time than the lower parts, and thus forms a roof, which sooner or later falls in, as in the above mentioned five cases. The downward movement of the clay may be compared with that of a glacier, but is incomparably slower; and this movement accounts for a singular fact, namely that the much elongated flints which are embedded in the chalk in a nearly horizontal position, are commonly found standing nearly or quite upright in the red clay. This fact is so common that the workmen assured me that this was their natural position. I roughly measured one which stood vertically, and it was of the same length and of the same relative thickness as one of my arms. These elongated flints must get placed in their upright position, been left on the surface and forms a bed of stiff red clay, full of flints, and generally from 6 to 14 feet in thickness. Over the red clay, wherever the land has long remained as pasture, there is a layer a few inches in thickness, of dark-coloured vegetable mould.

A quantity of broken chalk was spread, on December 20, 1842, over a part of a field near my house, which had existed as pasture certainly for 30, probably for twice or thrice as many years. The chalk was laid on the land for the sake of observing at some future period to what depth it would become buried. At the end of November, 1871, that is after an interval of 29 years, a trench was dug across this part of the field; and a line of white nodules could be traced on both sides of the trench, at a depth of 7 inches from the surface. The mould, therefore, (excluding the turf) had

on the same principle that a trunk of a tree left on a glacier assumes a position parallel to the line of motion. The flints in the clay which form almost half its bulk, are very often broken, though not rolled or abraded; and this may be accounted for by their mutual pressure, whilst the whole mass is subsiding. I may add that the chalk here appears to have been originally covered in parts by a thin bed of fine sand with some perfectly rounded flint pebbles, probably of Tertiary age; for such sand often partly fills up the deeper pits or cavities in the chalk.

here been thrown up at an average rate of ·22 inches per year. Beneath the line of chalk nodules there was in parts hardly any fine earth free of flints, while in other parts there was a layer, 2¼ inches in thickness. In this latter case the mould was altogether 9¼ inches thick; and in one such spot a nodule of chalk and a smooth flint pebble, both of which must have been left at some former time on the surface, were found at this depth. At from 11 to 12 inches beneath the surface, the undisturbed reddish clay, full of flints, extended. The appearance of the above nodules of chalk surprised me much at first, as they closely resembled water-worn pebbles, whereas the freshly-broken fragments had been angular. But on examining the nodules with a lens, they no longer appeared water-worn, for their surfaces were pitted through unequal corrosion, and minute, sharp points, formed of broken fossil shells, projected from them. It was evident that the corners of the original fragments of chalk had been wholly dissolved, from presenting a large surface to the carbonic acid dissolved in the rain-water and to that generated in soil containing vegetable matter, as well as to the humus-acids.* The projecting corners would also, relatively to the other parts, have been embraced by a larger number of living rootlets; and these have the power of even attacking marble, as Sachs has shown. Thus, in the course of 29 years, buried angular fragments of chalk had been converted into well-rounded nodules.

Another part of this same field was mossy, and as it was thought that sifted coal-cinders would improve the pasture, a thick layer was spread over this part either in 1842 or 1843, and another layer some years afterwards. In 1871 a trench was here dug, and many cinders lay in a line at a depth of 7 inches beneath the surface, with another line at a depth of 5½ inches parallel to the one beneath. In another part of this field, which had formerly existed as a separate one, and which it was believed had been pasture-land for more than a century, trenches were dug to see how thick the vegetable mould was. By chance the first trench was made at a spot where at some former period,

* S. W. Johnson, 'How Crops Feed,' 1870, p. 139.

certainly more than forty years before, a large hole had been filled up with coarse red clay, flints, fragments of chalk, and gravel; and here the fine vegetable mould was only from $4\frac{1}{8}$ to $4\frac{3}{8}$ inches in thickness. In another and undisturbed place, the mould varied much in thickness, namely from $6\frac{1}{2}$ to $8\frac{1}{2}$ inches; beneath which a few small fragments of brick were found in one place. From these several cases, it would appear that during the last 29 years mould has been heaped on the surface at an average annual rate of from ·2 to ·22 of an inch. But in this district when a ploughed field is first laid down in grass, the mould accumulates at a much slower rate. The rate, also, must become very much slower after a bed of mould, several inches in thickness, has been formed; for the worms then live chiefly near the surface, and burrow down to a greater depth so as to bring up fresh earth from below, only during the winter when the weather is very cold (at which time worms were found in this field at a depth of 26 inches) and during summer, when the weather is very dry.

A field, which adjoins the one just described, slopes in one part rather steeply (viz., at from 10° to 15°); this part was last ploughed in 1841, was then harrowed and left to become pasture-land. For several years it was clothed with an extremely scant vegetation, and was so thickly covered with small and large flints (some of them half as large as a child's head) that the field was always called by my sons "the stony field." When they ran down the slope the stones clattered together. I remember doubting whether I should live to see these larger flints covered with vegetable mould and turf. But the smaller stones disappeared before many years had elapsed, as did every one of the larger ones after a time; so that after thirty years (1871) a horse could gallop over the compact turf from one end of the field to the other, and not strike a single stone with his shoes. To anyone who remembered the appearance of the field in 1842, the transformation was wonderful. This was certainly the work of the worms, for though castings were not frequent for several years, yet some were thrown up month after month, and these gradually increased in numbers as the pasture improved. In the year 1871 a trench was dug on the above slope, and the blades of grass were cut off close to the roots, so that the thickness of the turf and of the vegetable mould could be measured accurately. The turf was rather less than half an inch, and the mould, which did not contain any stones, 2½ inches in thickness. Beneath this lay coarse clayey earth full of flints, like that in any of the neighbouring ploughed fields. This coarse earth easily fell apart from the overlying mould when a spit was lifted up. The average rate of accumulation of the mould during the whole thirty years was only ·083 inch per year (i.e., nearly one inch in twelve years); but the rate must have been much slower at first, and afterwards considerably quicker.

The transformation in the appearance of this field, which had been effected beneath my eyes, was afterwards rendered the more striking, when I examined in Knole Park a dense forest of lofty beech-trees, beneath which nothing grew. Here the ground was thickly strewed with large naked stones, and worm-castings were almost wholly absent. Obscure lines and irregularities on the surface indicated that the land had been cultivated some centuries ago. It is probable that a thick wood of young beech-trees sprung up so quickly, that time enough was not allowed for worms to cover up the stones with their castings, before the site became unfitted for their existence. Anyhow the contrast between the state of the now miscalled "stony field," well stocked with worms, and the present state of the ground beneath the old beech-trees in Knole Park, where worms appeared to be absent, was striking.

A narrow path running across part of my lawn was paved in 1843 with small flagstones, set edgeways; but worms threw up many castings and weeds grew thickly between them. During several years the path was weeded and swept; but ultimately the weeds and worms prevailed, and the gardener ceased to sweep, merely mowing off the weeds, as often as the lawn was mowed. The path soon became almost covered up, and after several years no trace of it was left. On removing, in 1877, the thin overlying layer of turf, the small flag-stones, all in their proper places, were found covered by an inch of fine mould.

Two recently published accounts of substances strewed on the surface of pasture-land, having become buried through the action of worms, may be here noticed. The Rev. H. C. Key had a ditch cut in a field, over which coal-ashes had been spread, as it was believed, eighteen years before; and on the clean-cut perpendicular sides of the ditch, at a depth of at least seven inches, there could be seen, for a length of 60 yards, "a distinct, very "even, narrow line of coal-ashes, mixed with "small coal, perfectly parallel with the top-sward."* This parallelism and the length of the section gives interest to the case. Secondly, Mr. Dancer states† that crushed bones had been thickly strewed over a field; and "some years "afterwards" these were found "several inches "below the surface, at a uniform depth." Worms appear to act in the same manner in New Zealand as in Europe; for Professor J.

* 'Nature,' November 1877, p. 28.

† 'Proc. Phil. Soc.' of Manchester, 1877, p. 247.

von Haast has described * a section near the coast, consisting of mica-schist, "covered by "5 or 6 feet of loess, above which about 12 "inches of vegetable soil had accumulated." Between the loess and the mould there was a layer from 3 to 6 inches in thickness, consisting of "cores, implements, flakes, and "chips, all manufactured from hard basaltic "rock." It is therefore probable that the aborigines, at some former period, had left these objects on the surface, and that they had afterwards been slowly covered up by the castings of worms.

Farmers in England are well aware that objects of all kinds, left on the surface of pasture-land, after a time disappear, or, as they say, work themselves downwards. How powdered lime, cinders, and heavy stones, can work down, and at the same rate, through the matted roots of a grass-covered surface, is a question which has probably never occurred to them.†

* 'Trans. of the New Zealand Institute,' vol. xii., 1880, p. 152.

† Mr. Lindsay Carnagie, in a letter (June 1838) to Sir C. Lyell, remarks that Scotch farmers are afraid of putting lime on ploughed land until just before it is laid down for pasture, from a belief that it has some tendency to sink. He adds: "Some

The Sinking of great Stones through the Action of Worms.—When a stone of large size and of irregular shape is left on the surface of the ground, it rests, of course, on the more protuberant parts; but worms soon fill up with their castings all the hollow spaces on the lower side; for, as Hensen remarks, they like the shelter of stones. As soon as the hollows are filled up, the worms eject the earth which they have swallowed beyond the circumference of the stones; and thus the surface of the ground is raised all round the stone. As the burrows excavated directly beneath the stone after a time collapse, the stone sinks a little.* Hence it is, that boulders which at some ancient

---

years since, in autumn, I laid lime on an oat-stubble and ploughed it down; thus bringing it into immediate contact with the dead vegetable matter, and securing its thorough mixture through the means of all the subsequent operations of fallow. In consequence of the above prejudice, I was considered to have committed a great fault; but the result was eminently successful, and the practice was *partially* followed. By means of Mr. Darwin's observations, I think the prejudice will be removed."

* This conclusion, which, as we shall immediately see, is fully justified, is of some little importance, as the so-called bench-stones, which surveyors fix in the ground as a record of their levels, may in time become false standards. My son Horace intends at some future period to ascertain how far this has occurred.

period have rolled down from a rocky mountain or cliff on to a meadow at its base, are always somewhat embedded in the soil; and, when removed, leave an exact impression of their lower surfaces in the under-lying fine mould. If, however, a boulder is of such huge dimensions, that the earth beneath is kept dry, such earth will not be inhabited by worms, and the boulder will not sink into the ground.

A lime-kiln formerly stood in a grass-field near Leith Hill Place in Surrey, and was pulled down 35 years before my visit; all the loose rubbish had been carted away, excepting three large stones of quartzose sandstone, which it was thought might hereafter be of some use. An old workman remembered that they had been left on a bare surface of broken bricks and mortar, close to the foundations of the kiln; but the whole surrounding surface is now covered with turf and mould. The two largest of these stones had never since been moved; nor could this easily have been done, as, when I had them removed, it was the work of two men with levers. One of these stones, and not the largest, was 64 inches long, 17 inches broad, and from 9 to 10 inches in thickness. Its lower surface was somewhat protuberant in the middle; and this part still rested on broken bricks and mortar, showing the truth of the old workman's account. Beneath the brick rubbish the natural sandy soil, full of fragments of sandstone was found; and this could have yielded very little, if at all, to the weight of the stone, as might have been expected if the sub-soil had been clay. The surface of the field, for a distance of about 9 inches round the stone, gradually sloped up to it, and close to the stone stood in most places about 4 inches above the surrounding ground. The base of the stone was buried from 1 to 2 inches beneath the general level, and the upper surface projected about 8 inches above this level, or about 4 inches above the sloping border of turf. After the removal of the stone it became evident that one of its pointed ends must at first have stood clear above the ground by some inches, but its upper surface was now on a level with the surrounding turf. When the stone was removed, an exact cast of its lower side, forming a shallow crateriform hollow, was left, the inner surface of which consisted of fine black mould, excepting where the more protuberant parts rested on the brick-rubbish. A transverse section of this stone, together with its bed, drawn from measurements made after it had been displaced, is here given on a scale of ½ inch to a foot (Fig. 6). The turf-covered border which

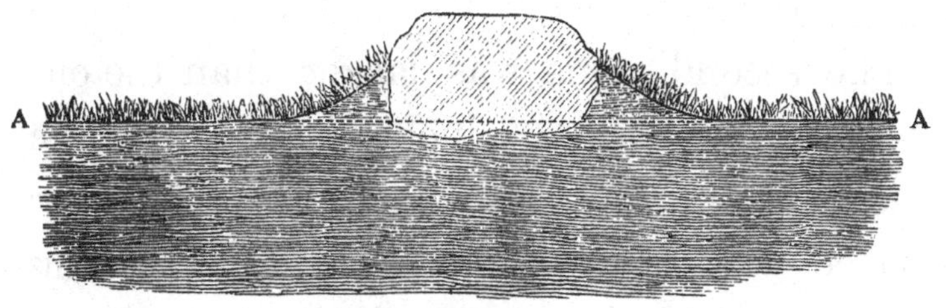

Fig. 6.

Transverse section across a large stone, which had lain on a grass-field for 35 years. A A, general level of the field. The underlying brick rubbish has not been represented. Scale ½ inch to one foot.

sloped up to the stone, consisted of fine vegetable mould, in one part 7 inches in thickness. This evidently consisted of worm-castings, several of which had been recently ejected. The whole stone had sunk in the thirty-five years, as far as I could judge, about 1½ inch; and this must have been due to the brick-rubbish beneath the more protuberant parts having been undermined by worms. At this rate the upper surface of the stone, if it had been left undisturbed, would have sunk to the general level of the field in 247 years; but before this could have occurred, some earth would have been washed down by heavy rain from the castings on the raised border of turf over the upper surface of the stone.

The second stone was larger than the one just described, viz., 67 inches in length, 39 in breadth, and 15 in thickness. The lower surface was nearly flat, so that the worms must soon have been compelled to eject their castings beyond its circumference. The stone as a whole had sunk about 2 inches into the ground. At this rate it would have required 262 years for its upper surface to have sunk to the general level of the field. The upwardly sloping, turf-covered border round the stone was broader than in the last case, viz., from 14 to 16 inches; and why this should be so, I could see no reason. In most parts this border was not so high as in the last case, viz., from 2 to 2½ inches, but in one place it was as much as $5\frac{1}{2}$. Its average height close to the stone was probably about 3 inches, and it thinned out to nothing. If so, a layer of fine earth, 15 inches in breadth and $1\frac{1}{2}$ inch in average thickness, of sufficient length to surround the whole of the much elongated slab, must have been brought up by the worms in chief part from beneath the stone in the course of 35 years. This amount would be amply sufficient to account for its having sunk about 2 inches into the ground; more especially if we bear in mind that a good deal of the finest earth would have been washed by heavy rain from the castings ejected on the sloping border down to the level of the field. Some fresh castings were seen close to the stone. Nevertheless, on digging a large hole to a depth of 18 inches where the stone had lain, only two worms and a few burrows were seen, although the soil was damp and seemed favourable for worms. There were some large colonies of ants beneath the stone, and possibly since their establishment the worms had decreased in number.

The third stone was only about half as

large as the others; and two strong boys could together have rolled it over. I have no doubt that it had been rolled over at a moderately recent time, for it now lay at some distance from the two other stones at the bottom of a little adjoining slope. It rested also on fine earth, instead of partly on brick-rubbish. In agreement with this conclusion, the raised surrounding border of turf was only 1 inch high in some parts, and 2 inches in other parts. There were no colonies of ants beneath this stone, and on digging a hole where it had lain, several burrows and worms were found.

At Stonehenge, some of the outer Druidical stones are now prostrate, having fallen at a remote but unknown period; and these have become buried to a moderate depth in the ground. They are surrounded by sloping borders of turf, on which recent castings were seen. Close to one of these fallen stones, which was 17 ft. long, 6 ft. broad, and 28½ inches thick, a hole was dug; and here the vegetable mould was at least 9½ inches in thickness. At this depth a flint was found, and a little higher up on one side of the hole a fragment of glass. The base of the stone lay about 9½ inches beneath the level of the surrounding ground, and its upper surface 19 inches above the ground.

A hole was also dug close to a second huge stone, which in falling had broken into two pieces; and this must have happened long ago, judging from the weathered aspect of the fractured ends. The base was buried to a depth of 10 inches, as was ascertained by driving an iron skewer horizontally into the ground beneath it. The vegetable mould forming the turf-covered sloping border round the stone, on which many castings had recently been ejected, was 10 inches in thickness; and most of this mould must have been brought up by worms from beneath its base. At a distance of 8 yards from the stone, the mould was only 5½ inches in thickness (with a piece of tobacco pipe at a depth of 4 inches), and this rested on broken flint and chalk which could not have easily yielded to the pressure or weight of the stone.

A straight rod was fixed horizontally (by the aid of a spirit-level) across a third fallen stone, which was 7 feet 9 inches long; and the contour of the projecting parts and of the adjoining ground, which was not quite level, was thus ascertained, as shown in the accompanying diagram (Fig. 7) on a scale of

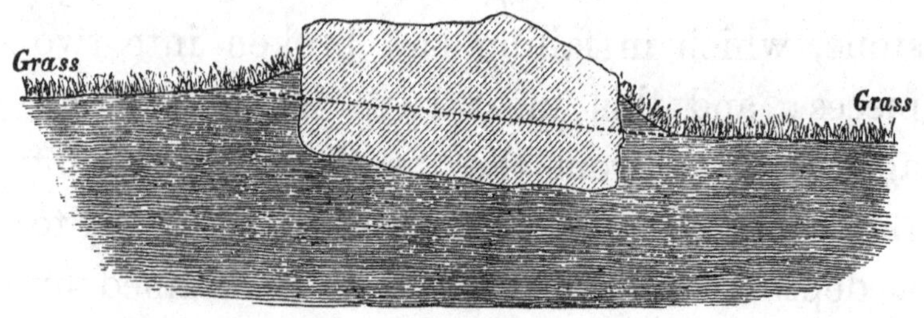

Fig. 7.

Section through one of the fallen Druidical stones at Stonehenge, showing how much it had sunk into the ground. Scale ½ inch to 1 foot.

½ inch to a foot. The turf-covered border sloped up to the stone on one side to a height of 4 inches, and on the opposite side to only 2½ inches above the general level. A hole was dug on the eastern side, and the base of the stone was here found to lie at a depth of 4 inches beneath the general level of the ground, and of 8 inches beneath the top of the sloping turf-covered border.

Sufficient evidence has now been given showing that small objects left on the surface of the land where worms abound soon get buried, and that large stones sink slowly downwards through the same means. Every step of the process could be followed, from the accidental deposition of a single casting on a small object lying loose on the surface, to its being entangled amidst the matted roots of the turf, and lastly to its being embedded in the mould at various depths beneath the surface. When the same field was re-examined after the interval of a few years, such objects were found at a greater depth than before. The straightness and regularity of the lines formed by the embedded objects, and their parallelism with the surface of the land, are the most striking features of the case; for this parallelism shows how equably the worms must have worked; the result being, however, partly the effect of the washing down of the fresh castings by rain. The specific gravity of the objects does not affect their rate of sinking, as could be seen by porous cinders, burnt marl, chalk and quartz pebbles, having all sunk to the same depth within the same time. Considering the nature of the substratum, which at Leith Hill

Place was sandy soil including many bits of rock, and at Stonehenge, chalk-rubble with broken flints; considering, also, the presence of the turf-covered sloping border of mould round the great fragments of stone at both these places, their sinking does not appear to have been sensibly aided by their weight, though this was considerable.*

On the number of worms which live within a given space.—We will now show, firstly, what a vast number of worms live unseen by us beneath our feet, and, secondly, the actual weight of the earth which they bring up to the surface within a given space and within a given time. Hensen, who has published so full and interesting an account of the habits of worms,† calculates, from the number which he found in a measured space, that there must exist 133,000 living worms in a hectare of

* Mr. R. Mallet remarks ('Quarterly Journal of Geolog. Soc. vol. xxxiii., 1877, p. 745) that "the extent to which the ground beneath the foundations of ponderous architectural structures, such as cathedral towers, has been known to become compressed, is as remarkable as it is instructive and curious. The amount of depression in some cases may be measured by feet." He instances the Tower of Pisa, but adds that it was founded on "dense clay."

† 'Zeitschrift für wissensch. Zoolog.' Bd. xxviii., 1877, p. 354

land, or 53,767 in an acre. This latter number of worms would weigh 356 pounds, taking Hensen's standard of the weight of a single worm, namely, one gram. It should, however, be noted that this calculation is founded on the numbers found in a garden, and Hensen believes that worms are here twice as numerous as in corn-fields. The above result, astonishing though it be, seems to me credible, judging from the number of worms which I have sometimes seen, and from the number daily destroyed by birds without the species being exterminated. Some barrels of bad ale were left on Mr. Miller's land,* in the hope of making vinegar, but the vinegar proved bad, and the barrels were upset. It should be premised that acetic acid is so deadly a poison to worms that Perrier found that a glass rod dipped into this acid and then into a considerable body of water in which worms were immersed, invariably killed them quickly. On the morning after the barrels had been upset, "the "heaps of worms which lay dead on the

* See Mr. Dancer's paper in 'Proc. Phil. Soc. of Manchester,' 1877, p. 248.

"ground were so amazing, that if Mr. Miller "had not seen them, he could not have "thought it possible for such numbers to "have existed in the space." As further evidence of the large number of worms which live in the ground, Hensen states that he found in a garden sixty-four open burrows in a space of 14½ square feet, that is, nine in 2 square feet. But the burrows are sometimes much more numerous, for when digging in a grass-field near Maer Hall, I found a cake of dry earth, as large as my two open hands, which was penetrated by seven burrows, as large as goose-quills.

Weight of the earth ejected from a single burrow, and from all the burrows within a given space.—With respect to the weight of the earth daily ejected by worms, Hensen found that it amounted, in the case of some worms which he kept in confinement, and which he appears to have fed with leaves, to only 0·5 gram, or less than 8 grains per diem. But a very much larger amount must be ejected by worms in their natural state, at the periods when they consume earth as food instead of leaves, and when they are making deep burrows. This is rendered almost certain by the following weights of the castings thrown up at the mouths of single burrows; the whole of which appeared to have been ejected within no long time, as was certainly the case in several instances. The castings were dried (excepting in one specified instance) by exposure during many days to the sun or before a hot fire.

WEIGHT OF THE CASTINGS ACCUMULATED AT THE MOUTH OF A SINGLE BURROW.

	Ounces.
(1.) Down, Kent (sub-soil red clay, full of flints, overlying the chalk). The largest casting which I could find on the flanks of a steep valley, the sub-soil being here shallow. In this one case, the casting was not well dried	3·98
(2.) Down.—Largest casting which I could find (consisting chiefly of calcareous matter), on extremely poor pasture land at the bottom of the valley mentioned under (1.)	3·87
(3.) Down.—A large casting, but not of unusual size, from a nearly level field, poor pasture, laid down in grass about 35 years before	1·22
(4.) Down.—Average weight of 11 not large castings ejected on a sloping surface on my lawn, after they had suffered some loss of weight from being exposed during a considerable length of time to rain ..	0·7
(5.) Near Nice in France.—Average weight of 12 castings of ordinary dimensions, collected by Dr. King on land which had not been mown for a long time and where worms abounded, viz., a lawn protected by shrubberies, near the sea; soil sandy and calcareous; these castings had been exposed for some time to rain, before being collected, and must have lost some weight by disintegration, but they still retained their form	1·37

WEIGHT OF THE CASTINGS ACCUMULATED AT THE MOUTH OF A SINGLE BURROW—*continued.*

	Ounces.
(6.) The heaviest of the above twelve castings ..	1·76
(7.) Lower Bengal.—Average weight of 22 castings, collected by Mr. J. Scott, and stated by him to have been thrown up in the course of one or two nights	1·24
(8.) The heaviest of the above 22 castings	2·09
(9.) Nilgiri Mountains, S. India; average weight of the 5 largest castings collected by Dr. King. They had been exposed to the rain of the last monsoon, and must have lost some weight	3·15
(10.) The heaviest of the above 5 castings	4·34

In this table we see that castings which had been ejected at the mouth of the same burrow, and which in most cases appeared fresh and always retained their vermiform configuration, generally exceeded an ounce in weight after being dried, and sometimes nearly equalled a quarter of a pound. On the Nilgiri mountains one casting even exceeded this latter weight. The largest castings in England were found on extremely poor pasture-land; and these, as far as I have seen, are generally larger than those on land producing a rich vegetation. It would appear that worms have to swallow a greater amount of earth on poor than on rich land, in order to obtain sufficient nutriment.

With respect to the tower-like castings

near Nice (Nos. 5 and 6 in the above table), Dr. King often found five or six of them on a square foot of surface; and these, judging from their average weight, would have weighed together 7½ ounces; so that the weight of those on a square yard would have been 4 lb. 3½ oz. Dr. King collected, near the close of the year 1872, all the castings which still retained their vermiform shape, whether broken down or not, from a square foot, in a place abounding with worms, on the summit of a bank, where no castings could have rolled down from above. These castings must have been ejected, as he judged from their appearance in reference to the rainy and dry periods near Nice, within the previous five or six months; they weighed 9½ oz., or 5 lb. 5½ oz. per square yard. After an interval of four months, Dr. King collected all the castings subsequently ejected on the same square foot of surface, and they weighed 2½ oz., or 1 lb. 6½ oz. per square yard. Therefore within about ten months, or we will say for safety's sake within a year, 12 oz. of castings were thrown up on this one square foot, or 6·75 pounds on the square yard; and this would give 14·58 tons per acre.

In a field at the bottom of a valley in the chalk (see No. 2 in the foregoing table), a square yard was measured at a spot where very large castings abounded; they appeared, however, almost equally numerous in a few other places. These castings, which retained perfectly their vermiform shape, were collected; and they weighed when partially dried, 1 lb. 13½ oz. This field had been rolled with a heavy agricultural roller fifty-two days before, and this would certainly have flattened every single casting on the land. The weather had been very dry for two or three weeks before the day of collection, so that not one casting appeared fresh or had been recently ejected. We may therefore assume that those which were weighed had been ejected within, we will say, forty-five days from the time when the field was rolled,—that is, one week short of the whole intervening period. I had examined the same part of the field shortly before it was rolled, and it then abounded with fresh castings. Worms do not work in dry weather during the summer, or in winter during severe frosts. If we assume that they work for only half the year—though this is too low an estimate—then the worms in this field would eject during the year, 83·87 pounds per square yard; or 18·12 tons per acre, assuming the whole surface to be equally productive in castings.

In the foregoing cases some of the necessary data had to be estimated, but in the two following cases the results are much more trustworthy. A lady, on whose accuracy I can implicitly rely, offered to collect during a year all the castings thrown up on two separate square yards, near Leith Hill Place, in Surrey. The amount collected was, however, somewhat less than that originally ejected by the worms; for, as I have repeatedly observed, a good deal of the finest earth is washed away, whenever castings are thrown up during or shortly before heavy rain. Small portions also adhered to the surrounding blades of grass, and it required too much time to detach every one of them. On sandy soil, as in the present instance, castings are liable to crumble after dry weather, and particles were thus often lost. The lady also occasionally left home for a week or two, and at such times the castings must have suffered still greater loss from exposure to the weather. These losses were, however, compensated to some extent by the collections having been made on one of the squares for four days, and on the other square for two days more than the year.

A space was selected (October 9th, 1870) on a broad, grass-covered terrace, which had been mowed and swept during many years. It faced the south, but was shaded during part of the day by trees. It had been formed at least a century ago by a great accumulation of small and large fragments of sandstone, together with some sandy earth, rammed down level. It is probable that it was at first protected by being covered with turf. This terrace, judging from the number of castings on it, was rather unfavourable for the existence of worms, in comparison with the neighbouring fields and an upper terrace. It was indeed surprising that as many worms could live here as were seen; for on digging a hole in this terrace, the black vegetable mould together with the turf was only four inches in thickness, beneath which lay the level surface of light-coloured sandy soil, with many fragments of sandstone. Before any castings were collected all the previously existing ones were carefully removed. The last day's collection was on October 14th, 1871. The castings were then well dried before a fire; and they weighed exactly $3\frac{1}{2}$ lbs. This would give for an acre of similar land 7·56 tons of dry earth annually ejected by worms.

The second square was marked on unenclosed common land, at a height of about 700 ft. above the sea, at some little distance from Leith Hill Tower. The surface was clothed with short, fine turf, and had never been disturbed by the hand of man. The spot selected appeared neither particularly favourable nor the reverse for worms; but I have often noticed that castings are especially abundant on common land, and this may, perhaps, be attributed to the poorness of the soil. The vegetable mould was here between three and four inches in thickness. As this spot was at some distance from the house where the lady lived, the castings were not collected at such short intervals of time as those on the terrace; consequently the loss of fine earth during rainy weather must have been greater in this than in the last case. The castings moreover were more sandy, and in collecting them during dry weather they sometimes crumbled into dust, and much was thus lost. Therefore it is certain that the worms brought up to the surface considerably more earth than that which was collected. The last collection was made on October 27th, 1871; i.e., 367 days after the square had been marked out and the surface cleared of all pre-existing castings. The collected castings, after being well dried, weighed 7·453 pounds; and this would give, for an acre of the same kind of land, 16·1 tons of annually ejected dry earth.

SUMMARY OF THE FOUR FOREGOING CASES.

(1.) Castings ejected near Nice within about a year, collected by Dr. King on a square foot of surface, calculated to yield per acre 14·58 tons.

(2.) Castings ejected during about 45 days on a square yard, in a field of poor pasture at the bottom of a large valley in the Chalk, calculated to yield annually per acre 18·12 tons.

(3.) Castings collected from a square yard on an old terrace at

Leith Hill Place, during 369 days, calculated to yield annually per acre 7·56 tons.

(4.) Castings collected from a square yard on Leith Hill Common during 367 days, calculated to yield annually per acre 16·1 tons.

The thickness of the layer of mould, which castings ejected during a year would form if uniformly spread out.—As we know from the two last cases in the above summary, the weight of the dried castings ejected by worms during a year on a square yard of surface, I wished to learn how thick a layer of ordinary mould this amount would form if spread uniformly over a square yard. The dry castings were therefore broken into small particles, and whilst being placed in a measure were well shaken and pressed down. Those collected on the Terrace amounted to 124·77 cubic inches; and this amount, if spread out over a square yard, would make a layer ·09612 inch in thickness. Those collected on the Common amounted to 197·56 cubic inches, and would make a similar layer ·1524 inch in thickness.

These thicknesses must, however, be corrected, for the triturated castings, after being well shaken down and pressed, did not make nearly so compact a mass as vegetable mould, though each separate particle was very compact. Yet mould is far from being compact, as is shown by the number of air-bubbles which rise up when the surface is flooded with water. It is moreover penetrated by many fine roots. To ascertain approximately by how much ordinary vegetable mould would be increased in bulk by being broken up into small particles and then dried, a thin oblong block of somewhat argillaceous mould (with the turf pared off) was measured before being broken up, was well dried and again measured. The drying caused it to shrink by $\frac{1}{7}$ of its original bulk, judging from exterior measurements alone. It was then triturated and partly reduced to powder, in the same manner as the castings had been treated, and its bulk now exceeded (notwithstanding shrinkage from drying) by $\frac{1}{16}$ that of the original block of damp mould. Therefore the above calculated thickness of the layer, formed by the castings from the Terrace, after being damped and spread over a square yard, would have to be reduced by $\frac{1}{16}$; and this will reduce the layer to ·09 of an inch, so that a layer ·9 inch in thickness would be formed in the course of ten years. On the same principle the castings from the Common would make in the course of a single year a layer ·1429 inch, or in the course of 10 years 1·429 inch, in thickness. We may say in round numbers that the thickness in the former case would amount to nearly 1 inch, and in the second case to nearly 1½ inch in 10 years.

In order to compare these results with those deduced from the rates at which small objects left on the surfaces of grass-fields become buried (as described in the early part of this chapter), we will give the following summary:—

SUMMARY OF THE THICKNESS OF THE MOULD ACCUMULATED OVER OBJECTS LEFT STREWED ON THE SURFACE, IN THE COURSE OF TEN YEARS.

The accumulation of mould during 14¾ years on the surface of a dry, sandy, grass-field near Maer Hall, amounted to 2·2 inches in 10 years.

The accumulation during 21½ years on a swampy field near Maer Hall, amounted to nearly 1·9 inch in 10 years.

The accumulation during 7 years on a very swampy field near Maer Hall amounted to 2·1 inches in 10 years.

The accumulation during 29 years, on good, argillaceous pasture-land over the Chalk at Down, amounted to 2·2 inches in 10 years.

The accumulation during 30 years on the side of a valley over the Chalk at Down, the soil being argillaceous, very poor, and only just converted into pasture (so that it was for some years unfavourable for worms), amounted to 0·83 inches in 10 years.

In these cases (excepting the last) it may be seen that the amount of earth brought to the surface during 10 years is somewhat greater than that calculated from the castings which were actually weighed. This excess may be partly accounted for by the loss which the weighed castings had previously undergone through being washed by rain, by the adhesion of particles to the blades of the surrounding grass, and by their crumbling when dry. Nor must we overlook other agencies which in all ordinary cases add to the amount of mould, and which would not be included in the castings that were collected, namely, the fine earth brought up to the surface by burrowing larvæ and insects, especially by ants. The earth brought up by moles generally has a somewhat different appearance from vegetable mould; but after a time would not be distinguishable from it. In dry countries, moreover, the wind plays an important part in carrying dust from one place to another, and even in England it must add to the mould on fields near great roads. But in our county these latter several agencies appear to be of quite subordinate importance in comparison with the action of worms.

We have no means of judging how great a weight of earth a single full-sized worm ejects during a year. Hensen estimates that 53,767 worms exist in an acre of land; but this is founded on the number found in gardens, and he believes that only about half as many live in corn-fields. How many live in old pasture land is unknown; but if we assume that half the above number, or 26,886 worms live on such land, then taking from the previous summary 15 tons as the weight of the castings annually thrown up on an acre of land, each worm must annually eject 20 ounces. A full-sized casting at the mouth of a single burrow often exceeds, as we have seen, an ounce in weight; and it is probable that worms eject more than 20 full-sized castings during a year. If they eject annually more than 20 ounces, we may infer that the worms which live in an acre of pasture land must be less than 26,886 in number.

Worms live chiefly in the superficial mould,

which is usually from 4 or 5 to 10 and even 12 inches in thickness; and it is this mould which passes over and over again through their bodies and is brought to the surface. But worms occasionally burrow into the subsoil to a much greater depth, and on such occasions they bring up earth from this greater depth; and this process has gone on for countless ages. Therefore the superficial layer of mould would ultimately attain, though at a slower and slower rate, a thickness equal to the depth to which worms ever burrow, were there not other opposing agencies at work which carry away to a lower level some of the finest earth which is continually being brought to the surface by worms. How great a thickness vegetable mould ever attains, I have not had good opportunities for observing; but in the next chapter, when we consider the burial of ancient buildings, some facts will be given on this head. In the two last chapters we shall see that the soil is actually increased, though only to a small degree, through the agency of worms; but their chief work is to sift the finer from the coarser particles, to mingle the whole with vegetable débris, and to saturate it with their intestinal secretions.

Finally, no one who considers the facts given in this chapter—on the burying of small objects and on the sinking of great stones left on the surface—on the vast number of worms which live within a moderate extent of ground—on the weight of the castings ejected from the mouth of the same burrow—on the weight of all the castings ejected within a known time on a measured space—will hereafter, as I believe, doubt that worms play an important part in nature.

CHAPTER IV.

THE PART WHICH WORMS HAVE PLAYED IN THE BURIAL OF ANCIENT BUILDINGS.

The accumulation of rubbish on the sites of great cities independent of the action of worms—The burial of a Roman villa at Abinger—The floors and walls penetrated by worms—Subsidence of a modern pavement—The buried pavement at Beaulieu Abbey—Roman villas at Chedworth and Brading—The remains of the Roman town at Silchester—The nature of the débris by which the remains are covered—The penetration of the tesselated floors and walls by worms—Subsidence of the floors—Thickness of the mould—The old Roman city of Wroxeter—Thickness of the mould—Depth of the foundations of some of the Buildings—Conclusion.

ARCHÆOLOGISTS are probably not aware how much they owe to worms for the preservation of many ancient objects. Coins, gold ornaments, stone implements, &c., if dropped on the surface of the ground, will infallibly be buried by the castings of worms in a few years, and will thus be safely preserved, until the land at some future time is turned up. For instance, many years ago a grass-field was ploughed on the northern side of the Severn, not far from Shrewsbury; and a surprising number of iron arrow-heads were found at the bottom of the furrows, which, as Mr. Blakeway, a local antiquary, believed, were relics of the battle of Shrewsbury in the year 1403, and no doubt had been originally left strewed on the battle-field. In the present chapter I shall show that not only implements, &c., are thus preserved, but that the floors and the remains of many ancient buildings in England have been buried so effectually, in large part through the action of worms, that they have been discovered in recent times solely through various accidents. The enormous beds of rubbish, several yards in thickness, which underlie many cities, such as Rome, Paris, and London, the lower ones being of great antiquity, are not here referred to, as they have not been in any way acted on by worms. When we consider how much matter is daily brought into a great city for building, fuel, clothing and food, and that in old times when the roads were bad and the work of the scavenger was neglected, a comparatively small amount was carried away, we may agree with Elie de Beaumont, who, in discussing this subject, says, "pour une voiture de matériaux "qui en sort, on y en fait entrer cent." * Nor should we overlook the effects of fires, the demolition of old buildings, and the removal of rubbish to the nearest vacant space.

Abinger, Surrey.—Late in the autumn of 1876, the ground in an old farm-yard at this place was dug to a depth of 2 to 2½ feet, and the workmen found various ancient remains. This led Mr. T. H. Farrer of Abinger Hall to have an adjoining ploughed field searched. On a trench being dug, a layer of concrete, still partly covered with tesseræ (small red tiles), and surrounded on two sides by broken-down walls, was soon discovered. It is believed † that this room formed part of the atrium or reception-room of a Roman villa. The walls of two or three other small rooms were afterwards discovered. Many fragments of pottery, other objects, and coins of several

* 'Leçons de Géologie pratique,' 1845, p. 142.

† A short account of this discovery was published in 'The Times' of January 2, 1878; and a fuller account in 'The Builder,' January 5, 1878.

Roman emperors, dating from 133 to 361, and perhaps to 375 A.D., were likewise found. Also a half-penny of George I., 1715. The presence of this latter coin seems an anomaly; but no doubt it was dropped on the ground during the last century, and since then there has been ample time for its burial under a considerable depth of the castings of worms. From the different dates of the Roman coins we may infer that the building was long inhabited. It was probably ruined and deserted 1400 or 1500 years ago.

I was present during the commencement of the excavations (August 20, 1877) and Mr. Farrer had two deep trenches dug at opposite ends of the atrium, so that I might examine the nature of the soil near the remains. The field sloped from east to west at an angle of about 7°; and one of the two trenches, shown in the accompanying section (Fig. 8) was at the upper or eastern end. The diagram is on a scale of $\frac{1}{20}$ of an inch to an inch; but the trench, which was between 4 and 5 feet broad, and in parts above 5 feet deep, has necessarily been reduced out of all proportion. The fine mould over the floor

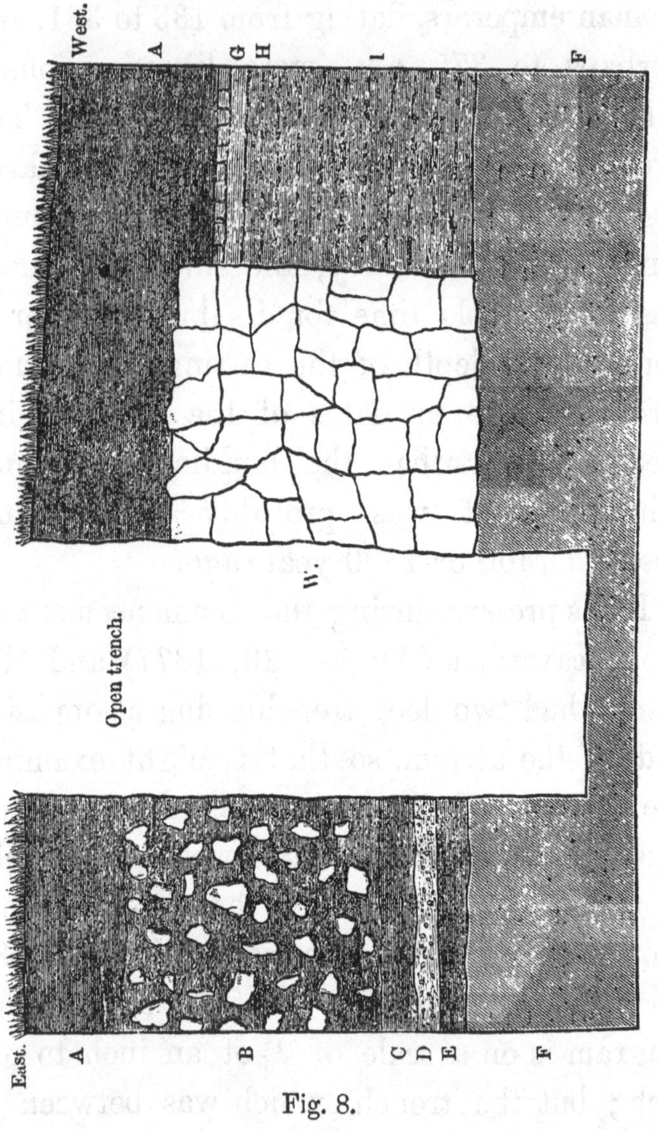

Fig. 8.

Section through the foundations of a buried Roman villa at Abinger. A A, vegetable mould; B, dark earth full of stones, 13 inches in thickness; C, black mould; D, broken mortar; E, black mould; F F, undisturbed sub-soil; G, tesseræ; H concrete; I, nature unknown; W, buried wall.

of the atrium varied in thickness from 11 to 16 inches; and on the side of the trench in the section was a little over 13 inches. After the mould had been removed, the floor appeared as a whole moderately level; but it sloped in parts at an angle of 1°, and in one place near the outside at as much as 8° 30′. The wall surrounding the pavement was built of rough stones, and was 23 inches in thickness where the trench was dug. Its broken summit was here 13 inches, but in another part 15 inches, beneath the surface of the field, being covered by this thickness of mould. In one spot, however, it rose to within 6 inches of the surface. On two sides of the room, where the junction of the concrete floor with the bounding walls could be carefully examined, there was no crack or separation. This trench afterwards proved to have been dug within an adjoining room (11 ft. by 11 ft. 6 in. in size), the existence of which was not even suspected whilst I was present.

On the side of the trench farthest from the buried wall (W), the mould varied from 9 to 14 inches in thickness; it rested on a mass (B)

23 inches thick of blackish earth, including many large stones. Beneath this was a thin bed of very black mould (C), then a layer of earth full of fragments of mortar (D), and then another thin bed (about 3 inches thick) (F) of very black mould, which rested on the undisturbed subsoil (F) of firm, yellowish, argillaceous sand. The 23-inch bed (B) was probably made ground, as this would have brought up the floor of the room to a level with that of the atrium. The two thin beds of black mould at the bottom of the trench evidently marked two former land-surfaces. Outside the walls of the northern room, many bones, ashes, oyster-shells, broken pottery and an entire pot were subsequently found at a depth of 16 inches beneath the surface.

The second trench was dug on the western or lower side of the villa: the mould was here only 6½ inches in thickness, and it rested on a mass of fine earth full of stones, broken tiles and fragments of mortar, 34 inches in thickness, beneath which was the undisturbed sand. Most of this earth had probably been washed down from the upper part of the field, and the fragments of stones, tiles, &c., must have come from the immediately adjoining ruins.

It appears at first sight a surprising fact that this field of light sandy soil should have been cultivated and ploughed during many years, and that not a vestige of these buildings should have been discovered. No one even suspected that the remains of a Roman villa lay hidden close beneath the surface. But the fact is less surprising when it is known that the field, as the bailiff believed, had never been ploughed to a greater depth than 4 inches. It is certain that when the land was first ploughed, the pavement and the surrounding broken walls must have been covered by at least 4 inches of soil, for otherwise the rotten concrete floor would have been scored by the ploughshare, the tesseræ torn up, and the tops of the old walls knocked down.

When the concrete and tesseræ were first cleared over a space of 14 by 9 ft., the floor which was coated with trodden-down earth exhibited no signs of having been penetrated by worms; and although the overlying fine mould closely resembled that which in many places has certainly been accumulated by worms, yet it seemed hardly possible that this mould could have been brought up by worms from beneath the apparently sound floor. It seemed also extremely improbable that the thick walls, surrounding the room and still united to the concrete, had been undermined by worms, and had thus been caused to sink, being afterwards covered up by their castings. I therefore at first concluded that all the fine mould above the ruins had been washed down from the upper parts of the field; but we shall soon see that this conclusion was certainly erroneous, though much fine earth is known to be washed down from the upper part of the field in its present ploughed state during heavy rains.

Although the concrete floor did not at first appear to have been anywhere penetrated by worms, yet by the next morning little cakes of the trodden-down earth had been lifted up by worms over the mouths of seven burrows, which passed through the softer parts of the naked concrete, or between the interstices of the tesseræ. On the third morning twenty-five burrows were counted;

and by suddenly lifting up the little cakes of earth, four worms were seen in the act of quickly retreating. Two castings were thrown up during the third night on the floor, and these were of large size. The season was not favourable for the full activity of worms, and the weather had lately been hot and dry, so that most of the worms now lived at a considerable depth. In digging the two trenches many open burrows and some worms were encountered at between 30 and 40 inches beneath the surface; but at a greater depth they became rare. One worm, however, was cut through at 48½, and another at 51½ inches beneath the surface. A fresh humus-lined burrow was also met with at a depth of 57 and another at 65½ inches. At greater depths than this, neither burrows nor worms were seen.

As I wished to learn how many worms lived beneath the floor of the atrium—a space of about 14 by 9 feet—Mr. Farrer was so kind as to make observations for me, during the next seven weeks, by which time the worms in the surrounding country were in full activity, and were working near the surface. It is very improbable that worms should have migrated from the adjoining field into the small space of the atrium, after the superficial mould in which they prefer to live, had been removed. We may therefore conclude that the burrows and the castings which were seen here during the ensuing seven weeks were the work of the former inhabitants of the space. I will now give a few extracts from Mr. Farrer's notes.

Aug. 26th, 1877; that is, five days after the floor had been cleared. On the previous night there had been some heavy rain, which washed the surface clean, and now the mouths of forty burrows were counted. Parts of the concrete were seen to be solid, and had never been penetrated by worms, and here the rain-water lodged.

Sept. 5th.—Tracks of worms, made during the previous night, could be seen on the surface of the floor, and five or six vermiform castings had been thrown up. These were defaced.

Sept. 12th.—During the last six days, the worms have not been active, though many castings have been ejected in the neighbouring fields; but on this day the earth was a little raised over the mouths of the burrows, or castings were ejected, at ten fresh points. These were defaced. It should be understood that when a fresh burrow is spoken of, this generally means only that an old burrow has been re-opened. Mr. Farrer was repeatedly struck with the pertinacity with which the worms re-opened their old burrows, even when no earth was ejected from them. I have often observed the same fact, and generally the mouths of the burrows are protected by an accumulation of pebbles, sticks or leaves. Mr. Farrer likewise observed that the worms living beneath the floor of the atrium often collected coarse grains of sand, and such little stones as they could find, round the mouths of their burrows.

Sept. 13th; soft wet weather. The mouths of the burrows were re-opened, or castings were ejected, at 31 points; these were all defaced.

Sept. 14th; 34 fresh holes or castings all defaced.

Sept. 15th; 44 fresh holes, only 5 castings; all defaced.

Sept. 18th; 43 fresh holes, 8 castings; all defaced.

The number of castings on the surrounding fields was now very large.

Sept. 19th; 40 holes, 8 castings; all defaced.

Sept. 22nd; 43 holes, only a few fresh castings; all defaced.

Sept. 23rd; 44 holes, 8 castings.

Sept. 25th; 50 holes, no record of the number of castings.

Oct. 13th; 61 holes, no record of the number of castings.

After an interval of three years, Mr. Farrer, at my request, again looked at the concrete floor, and found the worms still at work.

Knowing what great muscular power worms possess, and seeing how soft the concrete was in many parts, I was not surprised at its having been penetrated by their burrows; but it is a more surprising fact that the mortar between the rough stones of the thick walls, surrounding the rooms, was found by Mr. Farrer to have been penetrated by worms. On August 26th, that is, five days after the ruins had been exposed, he observed four open burrows on the broken summit of the eastern wall (W in Fig. 8); and, on September 15th, other burrows similarly situated were seen. It should also be noted that in the perpendicular side of the trench (which was much deeper than is represented in Fig. 8) three recent burrows were seen, which ran obliquely far down beneath the base of the old wall.

We thus see that many worms lived beneath the floor and the walls of the atrium at the time when the excavations were made; and that they afterwards almost daily brought up earth to the surface from a considerable depth. There is not the slightest reason to doubt that worms have acted in this manner ever since the period when the concrete was sufficiently decayed to allow them to penetrate it; and even before that period they would have lived beneath the floor, as soon as it became pervious to rain, so that the soil beneath was kept damp. The floor and the walls must therefore have been continually undermined; and fine earth must have been heaped on them during many centuries, perhaps for a thousand years. If the burrows beneath the floor and walls, which it is probable were formerly as numerous as they now are, had not collapsed in the course of time in the manner formerly explained, the underlying earth would have been riddled with passages like a sponge; and as this was not the case, we may feel sure that they have collapsed. The inevitable result of such collapsing during successive centuries, will have been the slow subsidence of the floor and of the walls, and their burial beneath the accumulated worm-castings. The subsidence of a floor, whilst it still remains nearly horizontal, may at first appear improbable; but the case presents no more real difficulty than that of loose objects strewed on the surface of a field, which, as we have seen, become buried several inches beneath the surface in the course of a few years, though still forming a horizontal layer parallel to the surface. The burial of the paved and level path on my lawn, which took place under my own observation, is an analogous case. Even those parts of the concrete floor which the worms could not penetrate would almost certainly have been undermined, and would have sunk, like the great stones at Leith Hill Place and Stonehenge, for the soil would have been damp beneath them. But the rate of sinking of the different parts would not have been quite equal, and the floor was not quite level. The foundations of the boundary walls lie, as shown in the section, at a very small depth beneath the surface; they would therefore have tended to subside at nearly the same rate as the floor. But this would not have occurred if the foundations had been deep, as in the case of some other Roman ruins presently to be described.

Finally, we may infer that a large part of the fine vegetable mould, which covered the floor and the broken-down walls of this villa, in some places to a thickness of 16 inches, was brought up from below by worms. From facts hereafter to be given there can be no doubt that some of the finest earth thus brought up will have been washed down the sloping surface of the field during every heavy shower of rain. If this had not occurred a greater amount of mould would have accumulated over the ruins than that now present. But beside the castings of worms and some earth brought up by insects, and some accumulation of dust, much fine earth will have been washed over the ruins from the upper parts of the field, since it has been under cultivation; and from over the ruins to the lower parts of the slope; the present thickness of the mould being the resultant of these several agencies.

I may here append a modern instance of the sinking of a pavement, communicated to me in 1871 by Mr. Ramsay, Director of the Geological Survey of England. A passage without a roof, 7 feet in length by 3 feet 2 inches in width, led from his house into the garden, and was paved with slabs of Portland stone. Several of these slabs were 16 inches square, others larger, and some a little smaller. This pavement had subsided about 3 inches along the middle of the passage, and two inches on each side, as could be seen by the lines of cement by which the slabs had been originally joined to the walls. The pavement had thus become slightly concave along the middle; but there was no subsidence at the end close to the house. Mr. Ramsay could not account for this sinking, until he observed that castings of black mould were frequently ejected along the lines of junction between the slabs; and these castings were regularly swept away. The several lines of junction, including those with the lateral walls, were altogether 39 feet 2 inches in length. The pavement did not present the appearance of ever having been renewed, and the house was believed to have been built about eighty-seven years ago. Considering all these circumstances, Mr. Ramsay does not doubt that the earth brought up by the worms since the pavement was first laid down, or rather since the decay of the mortar allowed the worms to burrow through it, and therefore within a much shorter time than the eighty-seven years, has sufficed to cause the sinking of the pavement to the above amount, except close to the house, where the ground beneath would have been kept nearly dry.

Beaulieu Abbey, Hampshire.—This abbey was destroyed by Henry VIII., and there now remains only a portion of the southern aisle-wall. It is believed that the king had most of the stones carried away for building a castle; and it is certain that they have been removed. The position of the nave-transept was ascertained not long ago by the foundations having been found; and the place is now marked by stones let into the ground. Where the abbey formerly stood, there now extends a smooth grass-covered surface, which resembles in all respects the rest of the field. The guardian, a very old man, said the surface had never been levelled in his time. In the year 1853, the Duke of Buccleuch had three holes dug in the turf within a few yards of one another, at the western end of the nave; and the old tesselated pavement of the abbey was thus discovered. These holes were afterwards surrounded by brickwork, and protected by trap-doors, so that the pavement might be readily inspected and preserved. When my son William examined the place on January 5, 1872, he found that the pavement in the three holes lay at depths of $6\frac{3}{4}$, 10 and $11\frac{1}{2}$ inches beneath the surrounding turf-covered surface. The old guardian asserted that he was often forced to remove worm-castings from the pavement; and that he had done so about six months before. My son collected all from one of the holes, the area of which was 5·32 square feet, and they weighed 7·97 ounces. Assuming that this amount had accumulated in six months, the accumulation during a year on a square yard would be 1·68 pounds, which, though a large amount, is very small compared with what, as we have seen, is often ejected on fields and commons. When I visited the abbey on June 22, 1877, the old man said that he had cleared out the holes about a month before, but a good many castings had since been ejected. I suspect that he imagined that he swept the pavements oftener than he really did, for the conditions were in several respects very unfavourable for the accumulation of even a moderate amount of castings. The tiles are rather large, viz., about 5½ inches square, and the mortar between them was in most places sound, so that the worms were able to bring up earth from below only at certain points. The tiles rested on a bed of concrete, and the castings in consequence consisted in large part (viz., in the proportion of 19 to 33) of particles of mortar, grains of sand, little fragments of rock, bricks or tile; and such substances could hardly be agreeable, and certainly not nutritious, to worms.

My son dug holes in several places within the former walls of the abbey, at a distance of several yards from the above described bricked squares. He did not find any tiles, though these are known to occur in some other parts, but he came in one spot to concrete on which tiles had once rested. The fine mould beneath the turf on the sides of the several holes, varied in thickness from only 2 to 2¾ inches, and this rested on a layer from 8¾ to above 11 inches in thickness, consisting of fragments of mortar and stone-rubbish with the interstices compactly filled up with black mould. In the surrounding field, at a distance of 20 yards from the abbey, the fine vegetable mould was 11 inches thick.

We may conclude from these facts that when the abbey was destroyed and the stones removed, a layer of rubbish was left over the whole surface, and that as soon as the worms were able to penetrate the decayed concrete and the joints between the tiles, they slowly filled up the interstices in the overlying rubbish with their castings, which were afterwards accumulated to a thickness of nearly three inches over the whole surface. If we add to this latter amount the mould between the fragments of stones, some five or six inches of mould must have been brought up from beneath the concrete or tiles. The concrete or tiles will consequently have subsided to nearly this amount. The bases of the columns of the aisles are now buried beneath mould and turf. It is not probable that they can have been undermined by worms, for their foundations would no doubt have been laid at a considerable depth. If they have not subsided, the stones of which the columns were constructed must have been removed from beneath the former level of the floor.

Chedworth, Gloucestershire.—The remains of a large Roman villa were discovered here in 1866, on ground which had been covered with wood from time immemorial. No suspicion seems ever to have been entertained that ancient buildings lay buried here, until a gamekeeper, in digging for rabbits, encountered some remains.* But subsequently the tops of some stone walls were detected in parts of the wood, projecting a little above the surface of the ground. Most of the coins found here belonged to Constans (who died 350 A.D.) and the Constantine family. My sons Francis and Horace visited the place in November 1877, for the sake of ascertaining what part worms may have played in the burial of these extensive remains. But the circumstances were not favourable for this object, as the ruins are surrounded on three sides by rather steep banks, down which earth is washed during rainy weather. Moreover most of the old rooms have been covered with roofs, for the protection of the elegant tesselated pavements.

A few facts may, however, be given on the thickness of the soil over these ruins. Close outside the northern rooms there is a broken wall, the summit of which was covered by 5

* Several accounts of these ruins have been published; the best is by Mr. James Farrer in 'Proc. Soc. of Antiquaries of Scotland,' vol. vi., Part II., 1867, p. 278. Also J. W. Grover, 'Journal of the British Arch. Assoc.' June 1866. Professor Buckman has likewise published a pamphlet, 'Notes on the Roman Villa at Chedworth,' 2nd edit. 1873: Cirencester.

inches of black mould; and in a hole dug on the outer side of this wall, where the ground had never before been disturbed, black mould, full of stones, 26 inches in thickness, was found, resting on the undisturbed sub-soil of yellow clay. At a depth of 22 inches from the surface a pig's jaw and a fragment of a tile were found. When the excavations were first made, some large trees grew over the ruins; and the stump of one has been left directly over a party-wall near the bath room, for the sake of showing the thickness of the superincumbent soil, which was here 38 inches. In one small room, which, after being cleared out, had not been roofed over, my sons observed the hole of a worm passing through the rotten concrete, and a living worm was found within the concrete. In another open room worm-castings were seen on the floor, over which some earth had by this means been deposited, and here grass now grew.

Brading, Isle of Wight.—A fine Roman villa was discovered here in 1880; and by the end of October no less than 18 chambers had been more or less cleared. A coin dated

337 A.D. was found. My son William visited the place before the excavations were completed; and he informs me that most of the floors were at first covered with much rubbish and fallen stones, having their interstices completely filled up with mould, abounding, as the workmen said, with worms, above which there was mould without any stones. The whole mass was in most places from 3 to above 4 ft. in thickness. In one very large room the overlying earth was only 2 ft. 6 in. thick; and after this had been removed, so many castings were thrown up between the tiles that the surface had to be almost daily swept. Most of the floors were fairly level. The tops of the broken-down walls were covered in some places by only 4 or 5 inches of soil, so that they were occasionally struck by the plough, but in other places they were covered by from 13 to 18 inches of soil. It is not probable that these walls could have been undermined by worms and subsided, as they rested on a foundation of very hard red sand, into which worms could hardly burrow. The mortar, however, between the stones of the walls of

a hypocaust was found by my son to have been penetrated by many worm-burrows. The remains of this villa stand on land which slopes at an angle of about 3°; and the land appears to have been long cultivated. Therefore no doubt a considerable quantity of fine earth has been washed down from the upper parts of the field, and has largely aided in the burial of these remains.

Silchester, Hampshire.—The ruins of this small Roman town have been better preserved than any other remains of the kind in England. A broken wall, in most parts from 15 to 18 feet in height and about $1\frac{1}{2}$ mile in compass, now surrounds a space of about 100 acres of cultivated land, on which a farm-house and a church stand.* Formerly, when the weather was dry, the lines of the buried walls could be traced by the appearance of the crops; and recently very extensive excavations have been undertaken by the Duke of Wellington, under the superintendence of the late Rev. J. G. Joyce, by which means many large buildings have been

* These details are taken from the 'Penny Encyclopædia,' article, Hampshire.

discovered. Mr. Joyce made careful coloured sections, and measured the thickness of each bed of rubbish, whilst the excavations were in progress; and he has had the kindness to send me copies of several of them. When my sons Francis and Horace visited these ruins, he accompanied them, and added his notes to theirs.

Mr. Joyce estimates that the town was inhabited by the Romans for about three centuries; and no doubt much matter must have accumulated within the walls during this long period. It appears to have been destroyed by fire, and most of the stones used in the buildings have since been carried away. These circumstances are unfavourable for ascertaining the part which worms have played in the burial of the ruins; but as careful sections of the rubbish overlying an ancient town have seldom or never before been made in England, I will give copies of the most characteristic portions of some of those made by Mr. Joyce. They are of too great length to be here introduced entire.

An east and west section, 30 ft. in length, was made across a room in the Basilica, now called the Hall of the Merchants (Fig. 9). The hard concrete floor, still covered here and there with tesseræ, was found at 3 ft.

Fig. 9.

Section within a room in the Basilica at Silchester. Scale $\frac{1}{18}$.

beneath the surface of the field, which was here level. On the floor there were two large piles of charred wood, one alone of which is shown in the part of the section here given. This pile was covered by a thin white layer of decayed stucco or plaster, above which was a mass, presenting a singularly disturbed appearance, of broken tiles, mortar, rubbish and fine gravel, together 27 inches in thickness. Mr. Joyce believes that the gravel was used in making the mortar or concrete, which has since decayed, some of the lime probably having been dissolved. The disturbed state of the rubbish may have been due to its having been searched for building stones. This bed was capped by fine vegetable mould, 9 inches in thickness. From these facts we may conclude that the Hall was burnt down, and that much rubbish fell on the floor, through and from which the worms slowly brought up the mould, now forming the surface of the level field.

A section across the middle of another hall in the Basilica, 32 feet 6 inches in length, called the Œvarium, is shown in Fig. 10. It appears that we have here evidence of two fires, separated by an interval of time, during which the 6 inches of "mortar and concrete

Fig. 10.

Section within a hall in the Basilica at Silchester. Scale $\frac{1}{32}$.

with broken tiles" was accumulated. Beneath one of the layers of charred wood, a valuable relic, a bronze eagle, was found; and this shows that the soldiers must have deserted the place in a panic. Owing to the death of Mr. Joyce, I have not been able to ascertain beneath which of the two layers the eagle was found. The bed of rubble overlying the undisturbed gravel originally formed, as I suppose, the floor, for it stands on a level with that of a corridor, outside the walls of the Hall; but the corridor is not shown in the section as here given. The vegetable mould was 16 inches thick in the thickest part; and the depth from the surface of the field, clothed with herbage, to the undisturbed gravel, was 40 inches.

The section shown in Fig. 11 represents an excavation made in the middle of the town, and is here introduced because the bed of "rich "mould" attained, according to Mr. Joyce, the unusual thickness of 20 inches. Gravel lay at the depth of 48 inches from the surface; but it was not ascertained whether this was in its natural state, or had been brought here and had been rammed down, as occurs in some other places.

The section shown in Fig. 12 was taken in the centre of the Basilica, and though it was 5 feet in depth, the natural sub-soil was not

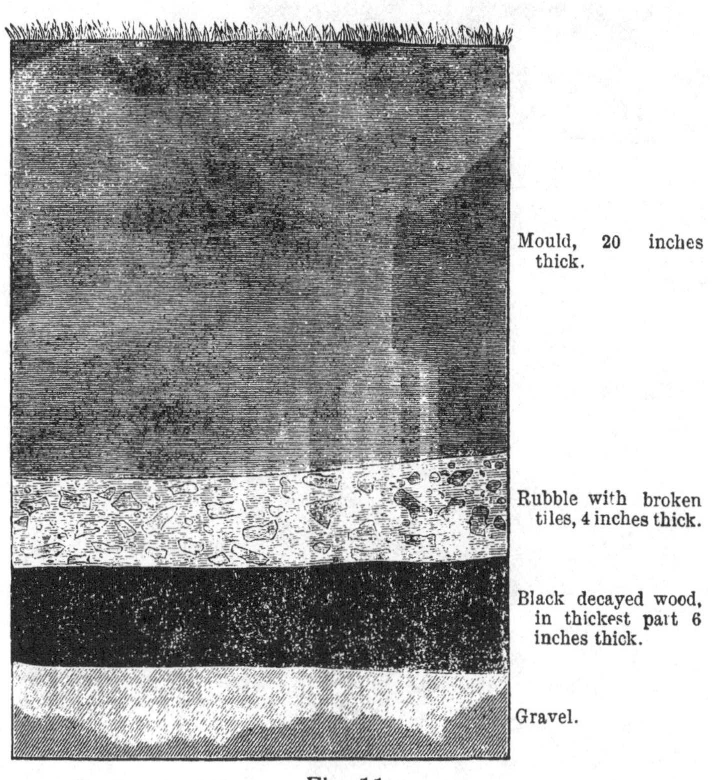

Fig. 11.

Section in a block of buildings in the middle of the town of Silchester.

reached. The bed marked "concrete" was probably at one time a floor; and the beds beneath seem to be the remnants of more ancient buildings. The vegetable mould was here only 9 inches thick. In some other sections, not copied, we likewise have

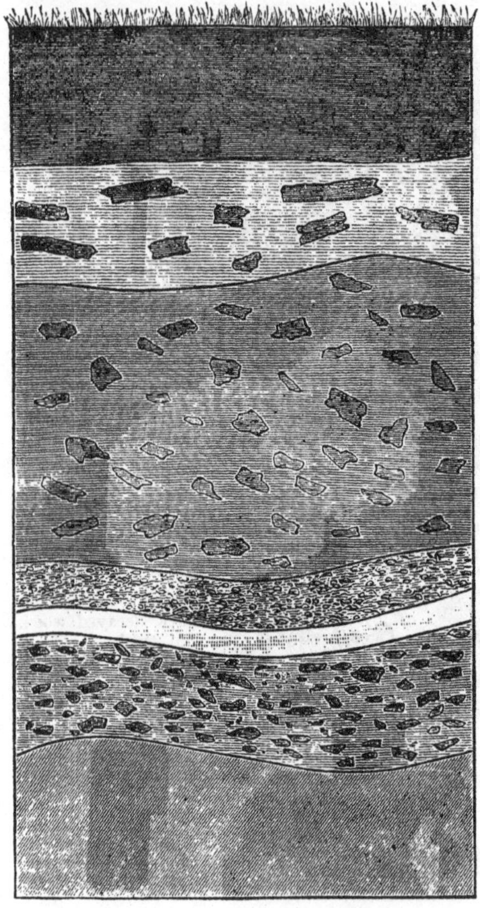

Fig. 12.

Section in the centre of the Basilica at Silchester.

evidence of buildings having been erected over the ruins of older ones. In one case there was a layer of yellow clay of very unequal thickness between two beds of débris, the lower one of which rested on a floor with tesseræ. The old broken walls appear sometimes to have been roughly cut down to a uniform level, so as to serve as the foundations of a temporary building; and Mr. Joyce suspects that some of these buildings were wattled sheds, plastered with clay, which would account for the above-mentioned layer of clay.

Turning now to the points which more immediately concern us. Worm-castings were observed on the floors of several of the rooms, in one of which the tesselation was unusually perfect. The tesseræ here consisted of little cubes of hard sandstone of about 1 inch, several of which were loose or projected slightly above the general level. One or occasionally two open worm-burrows were found beneath all the loose tesseræ. Worms have also penetrated the old walls of these ruins. A wall, which had just been exposed to view during the excavations then in progress, was examined; it was built of large flints, and was 18 inches in thickness.

It appeared sound, but when the soil was removed from beneath, the mortar in the lower part was found to be so much decayed that the flints fell apart from their own weight. Here, in the middle of the wall, at a depth of 29 inches beneath the old floor and of 49½ inches beneath the surface of the field, a living worm was found, and the mortar was penetrated by several burrows.

A second wall was exposed to view for the first time, and an open burrow was seen on its broken summit. By separating the flints this burrow was traced far down in the interior of the wall; but as some of the flints cohered firmly, the whole mass was disturbed in pulling down the wall, and the burrow could not be traced to the bottom. The foundations of a third wall, which appeared quite sound, lay at a depth of 4 feet beneath one of the floors, and of course at a considerably greater depth beneath the level of the ground. A large flint was wrenched out of the wall at about a foot from the base, and this required much force, as the mortar was sound; but behind the flint in the middle of the wall, the mortar was friable, and here there were worm-burrows. Mr. Joyce and my sons were surprised at the blackness of the mortar in this and in several other cases, and at the presence of mould in the interior of the walls. Some may have been placed there by the old builders instead of mortar; but we should remember that worms line their burrows with black humus. Moreover open spaces would almost certainly have been occasionally left between the large irregular flints; and these spaces, we may feel sure, would be filled up by the worms with their castings, as soon as they were able to penetrate the wall. Rain-water, oozing down the burrows would also carry fine dark-coloured particles into every crevice. Mr. Joyce was at first very sceptical about the amount of work which I attributed to worms; but he ends his notes with reference to the last-mentioned wall by saying, "This "case caused me more surprise and brought "more conviction to me than any other. I "should have said, and did say, that it was "quite impossible such a wall could have been "penetrated by earth-worms."

In almost all the rooms the pavement has

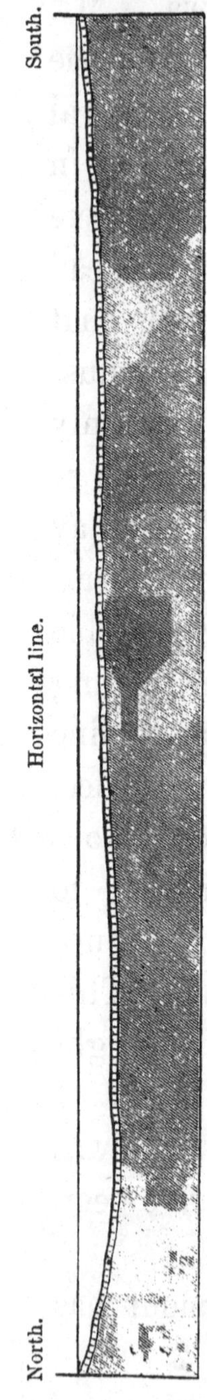

Fig. 13.

Section of the subsided floor of a room, paved with tesseræ, at Silchester. Scale $\frac{1}{40}$.

sunk considerably, especially towards the middle; and this is shown in the three following sections. The measurements were made by stretching a string tightly and horizontally over the floor. The section, Fig. 13, was taken from north to south across a room, 18 feet 4 inches in length, with a nearly perfect pavement, next to the "Red Wooden Hut." In the northern half, the subsidence amounted to $5\frac{3}{4}$ inches beneath the level of the floor as it now stands close to the walls; and it was greater in the northern than in the southern half; but, according to Mr. Joyce, the entire pavement has obviously subsided. In several places, the tesseræ appeared as if drawn a little away from the walls; whilst in other places they were still in close contact with them.

In Fig. 14, we see a section across the paved floor of the southern corridor or ambulatory of a quadrangle, in an excavation made near "The Spring." The floor is 7 feet 9 inches wide, and the broken-down walls now project only $\frac{3}{4}$ of an inch above its level. The field, which was in pasture, here sloped from north to south, at an angle of 3° 40′. The nature of the ground on each side of the corridor is shown in the section. It consisted of earth full of stones and other débris, capped with dark vegetable mould which was thicker on the lower or southern than on the northern side. The pavement was nearly level along lines parallel to the side-walls, but had sunk in the middle as much as $7\frac{3}{4}$ inches.

A small room at no great distance from that represented in Fig. 13, had been enlarged by the Roman occupier on the southern side, by an addition of 5 feet 4 inches in breadth. For this purpose the southern wall of the house had been pulled down, but the foundations of the old wall had been left buried at a little depth

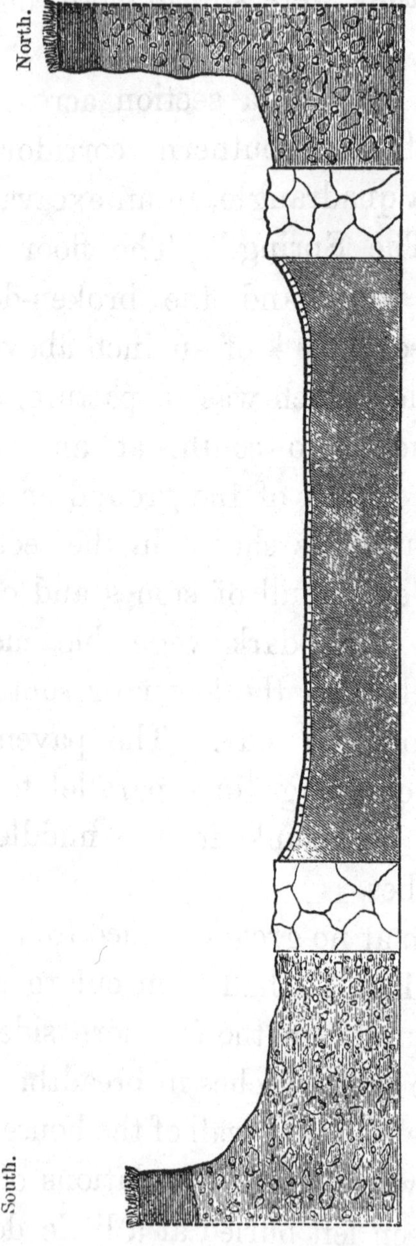

Fig. 14.

A north and south section through the subsided floor of a corridor, paved with tesseræ. Outside the broken-down bounding walls, the excavated ground on each side is shown for a short space. Nature of the ground beneath the tesseræ unknown. Silchester. Scale $\frac{1}{36}$.

beneath the pavement of the enlarged room. Mr. Joyce believes that this buried wall must have been built before the reign of Claudius II., who died 270, A.D. We see in the accompanying section, Fig. 15, that the tesselated pavement has subsided to a less degree over the buried wall than elsewhere; so that a slight convexity or protuberance here stretched in a straight line across the room. This led to a hole being dug, and the buried wall was thus discovered.

We see in these three sections, and in several others not given, that the old pavements have sunk or sagged considerably. Mr. Joyce formerly attributed this sinking solely to the slow settling of the ground. That there has been some settling is highly probable, and it may be seen in section 15 that the pavement for a width of 5 feet over the southern enlargement of the room, which must have been built on fresh ground, has sunk a little more than on the old northern side. But this sinking may possibly have had no connection with the enlargement of the room, for in Fig. 13, one half of the pavement has subsided more

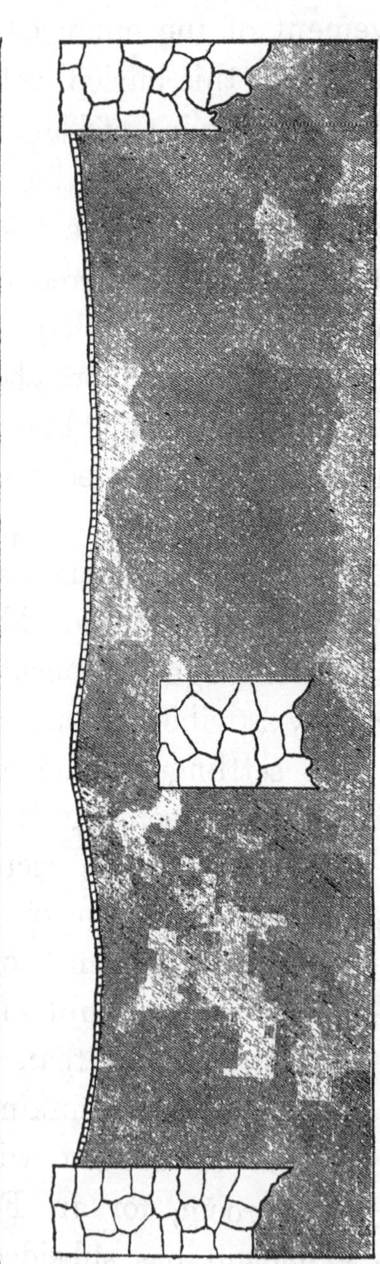

Fig. 15.

Section through the subsided floor, paved with tesseræ, and of the broken-down bounding walls of a room at Silchester, which had been formerly enlarged, with the foundations of the old wall left buried. Scale $\frac{1}{40}$.

than the other half without any assignable cause. In a bricked passage to Mr. Joyce's own house, laid down only about six years ago, the same kind of sinking has occurred as in the ancient buildings. Nevertheless it does not appear probable that the whole amount of sinking can be thus accounted for. The Roman builders excavated the ground to an unusual depth for the foundations of their walls, which were thick and solid; it is therefore hardly credible that they should have been careless about the solidity of the bed on which their tesselated and often ornamented pavements were laid. The sinking must, as it appears to me, be attributed in chief part to the pavement having been undermined by worms, which we know are still at work. Even Mr. Joyce at last admitted that this could not have failed to have produced a considerable effect. Thus also the large quantity of fine mould overlying the pavements can be accounted for, the presence of which would otherwise be inexplicable. My sons noticed that in one room in which the pavement had sagged very little, there was an unusually small amount of overlying mould.

As the foundations of the walls generally lie at a considerable depth, they will either have not subsided at all through the undermining action of worms, or they will have subsided much less than the floor. This latter result would follow from worms not often working deep down beneath the foundations; but more especially from the walls not yielding when penetrated by worms, whereas the successively formed burrows in a mass of earth, equal to one of the walls in depth and thickness, would have collapsed many times since the desertion of the ruins, and would consequently have shrunk or subsided. As the walls cannot have sunk much or at all, the immediately adjoining pavement from adhering to them will have been prevented from subsiding; and thus the present curvature of the pavement is intelligible.

The circumstance which has surprised me most with respect to Silchester is that during the many centuries which have elapsed since the old buildings were deserted, the vegetable mould has not accumulated over them to a greater thickness than that here observed. In most places it is only about 9 inches in thickness, but in some places 12 or even more inches. In Fig. 11, it is given as 20 inches, but this section was drawn by Mr. Joyce before his attention was particularly called to this subject. The land enclosed within the old walls is described as sloping slightly to the south; but there are parts which, according to Mr. Joyce, are nearly level, and it appears that the mould is here generally thicker than elsewhere. The surface slopes in other parts from west to east, and Mr. Joyce describes one floor as covered at the western end by rubbish and mould to a thickness of 28½ inches, and at the eastern end by a thickness of only 11½ inches. A very slight slope suffices to cause recent castings to flow downwards during heavy rain, and thus much earth will ultimately reach the neighbouring rills and streams and be carried away. By this means, the absence of very thick beds of mould over these ancient ruins may, as I believe, be explained. Moreover most of the land here has long been ploughed, and this would greatly aid the washing away of the finer earth during rainy weather.

The nature of the beds immediately beneath the vegetable mould in some of the sections is rather perplexing. We see, for instance, in the section of an excavation in a grass meadow (Fig. 14), which sloped from north to south at an angle of 3° 40′, that the mould on the upper side is only six inches and on the lower side nine inches in thickness. But this mould lies on a mass ($25\frac{1}{2}$ inches in thickness on the upper side) "of "dark brown mould," as described by Mr. Joyce, "thickly interspersed with small "pebbles and bits of tiles, which present a "corroded or worn appearance." The state of this dark-coloured earth is like that of a field which has long been ploughed, for the earth thus becomes intermingled with stones and fragments of all kinds which have been much exposed to the weather. If during the course of many centuries this grass meadow and the other now cultivated fields have been at times ploughed, and at other times left as pasture, the nature of the ground in the above section is rendered intelligible. For worms will continually have brought up fine earth from below, which will have been stirred up by the plough whenever the land was cultivated. But after a time a greater thickness of fine earth will thus have been accumulated than could be reached by the plough; and a bed like the $25\frac{1}{2}$-inch mass, in Fig. 14, will have been formed beneath the superficial mould, which latter will have been brought to the surface within more recent times, and have been well sifted by the worms.

Wroxeter, Shropshire.—The old Roman city of Uriconium was founded in the early part of the second century, if not before this date; and it was destroyed, according to Mr. Wright, probably between the middle of the fourth and fifth century. The inhabitants were massacred, and skeletons of women were found in the hypocausts. Before the year 1859, the sole remnant of the city above ground, was a portion of a massive wall about 20 ft. in height. The surrounding land undulates slightly, and has long been under cultivation. It had been noticed that the corn-crops ripened prematurely in certain narrow lines, and that the snow remained unmelted in certain places longer than in others.

These appearances led, as I was informed, to extensive excavations being undertaken. The foundations of many large buildings and several streets have thus been exposed to view. The space enclosed within the old walls is an irregular oval, about 1¾ mile in length. Many of the stones or bricks used in the buildings must have been carried away; but the hypocausts, baths, and other underground buildings were found tolerably perfect, being filled with stones, broken tiles, rubbish and soil. The old floors of various rooms were covered with rubble. As I was anxious to know how thick the mantle of mould and rubbish was, which had so long concealed these ruins, I applied to Dr. H. Johnson, who had superintended the excavations; and he, with the greatest kindness, twice visited the place to examine it in reference to my questions, and had many trenches dug in four fields which had hitherto been undisturbed. The results of his observations are given in the following Table. He also sent me specimens of the mould, and answered, as far as he could, all my questions.

MEASUREMENTS BY DR. H. JOHNSON OF THE THICKNESS OF THE VEGETABLE MOULD OVER THE ROMAN RUINS AT WROXETER.

Trenches dug in a field called "Old Works."

	Thickness of mould in inches.
1. At a depth of 36 inches undisturbed sand was reached	20
2. At a depth of 33 inches concrete was reached	21
3. " " 9 inches concrete was reached	9

Trenches dug in a field called "Shop "Leasows;" this is the highest field within the old walls, and slopes down from a sub-central point on all sides at about an angle of 2°.

	Thickness of mould in inches.
4. Summit of field, trench 45 inches deep ..	40
5. Close to summit of field, trench 36 inches deep	26
6. " " trench 28 inches deep	28
7. Near summit of field, trench 36 inches deep	24
8. " " trench at one end 39 inches deep; the mould here graduated into the underlying undisturbed sand, and its thickness is somewhat arbitrary. At the other end of the trench, a causeway was encountered at a depth of only 7 inches, and the mould was here only 7 inches thick ..	24
9. Trench close to the last, 28 inches in depth ..	15
10. Lower part of same field, trench 30 inches deep	15
11. " " trench 31 inches deep	17
12. " " trench 36 inches deep, at which depth undisturbed sand was reached	28

	Thickness of mould in inches.
13. In another part of same field, trench 9½ inches deep, stopped by concrete	9½
14. In another part of same field, trench 9 inches deep, stopped by concrete	9
15. In another part of the same field, trench 24 inches deep, when sand was reached ..	16
16. In another part of same field, trench 30 inches deep, when stones were reached ; at one end of the trench mould 12 inches, at the other end 14 inches thick	13

Small field between "Old Works" and "Shop Leasows," I believe nearly as high as the upper part of the latter field.

	Thickness of mould in inches.
17. Trench 26 inches deep	24
18. " 10 inches deep, and then came upon a causeway	10
19. Trench 34 inches deep	30
20. " 31 inches deep	31

Field on the western side of the space enclosed within the old walls.

	Thickness of mould in inches.
21. Trench 28 inches deep, when undisturbed sand was reached	16
22. Trench 29 inches deep, when undisturbed sand was reached	15
23. Trench 14 inches deep, and then came upon a building	14

Dr. Johnson distinguished as mould the earth which differed, more or less abruptly, in its dark colour and in its texture from the underlying sand or rubble. In the specimens sent to me, the mould resembled that which lies immediately beneath the turf in old pasture-land, excepting that it often contained small stones, too large to have passed through the bodies of worms. But the trenches above described were dug in fields, none of which were in pasture, and all had been long cultivated. Bearing in mind the remarks made in reference to Silchester on the effects of long-continued culture, combined with the action of worms in bringing up the finer particles to the surface, the mould, as so designated by Dr. Johnson, seems fairly well to deserve its name. Its thickness, where there was no causeway, floor or walls beneath, was greater than has been elsewhere observed, namely in many places above 2 ft., and in one spot above 3 ft. The mould was thickest on and close to the nearly level summit of the field called "Shop Leasows," and in a small adjoining field, which, as I believe, is of nearly the same height. One side of the former field slopes at an angle of rather above 2°, and I should have expected that the mould, from being washed down during heavy rain, would have been thicker in the lower than in the upper part; but this was not the case in two out of the three trenches here dug.

In many places, where streets ran beneath the surface, or where old buildings stood, the mould was only 8 inches in thickness; and Dr. Johnson was surprised that in ploughing the land, the ruins had never been struck by the plough as far as he had heard. He thinks that when the land was first cultivated the old walls were perhaps intentionally pulled down, and that hollow places were filled up. This may have been the case; but if after the desertion of the city the land was left for many centuries uncultivated, worms would have brought up enough fine earth to have covered the ruins completely; that is if they had subsided from having been undermined. The foundations of some of the walls, for instance those of the portion still standing about 20 feet above the ground, and those of the market-place, lie at the extraordinary depth of 14 feet; but it is highly improbable that the foundations were generally so deep. The mortar employed in the buildings must have been excellent, for it is still in parts extremely hard. Where-ever walls of any height have been exposed to view, they are, as Dr. Johnson believes, still perpendicular. The walls with such deep foundations cannot have been undermined by worms, and therefore cannot have subsided, as appears to have occurred at Abinger and Silchester. Hence it is very difficult to account for their being now completely covered with earth; but how much of this covering consists of vegetable mould and how much of rubble I do not know. The market-place, with the foundations at a depth of 14 feet, was covered up, as Dr. Johnson believes, by between 6 and 24 inches of earth. The tops of the broken-down walls of a caldarium or bath, 9 feet in depth, were likewise covered up with nearly 2 feet of earth. The summit of an arch, leading into an ash-pit 7 feet in depth, was covered up with not more than 8 inches of earth. Whenever a building which has not subsided is covered with earth, we must suppose, either that the upper layers of stone have been at some time carried away by man, or that earth has since been washed down during heavy rain, or blown down during storms, from the adjoining land; and this would be especially apt to occur where the land has long been cultivated. In the above cases the adjoining land is somewhat higher than the three specified sites, as far as I can judge by maps and from information given me by Dr. Johnson. If, however, a great pile of broken stones mortar, plaster, timber and ashes fell over the remains of any building, their disintegration in the course of time, and the sifting action of worms ,would ultimately conceal the whole beneath fine earth.

Conclusion.—The cases given in this chapter show that worms have played a considerable part in the burial and concealment of several Roman and other old buildings in England; but no doubt the washing down of soil from the neighbouring higher lands, and the deposition of dust, have together aided largely in the work of concealment. Dust would be apt to accumulate wherever old broken-down walls projected a little above the then existing surface and thus afforded some shelter. The floors of the old rooms, halls and passages have generally sunk, partly from the settling of the ground, but chiefly from having been undermined by worms; and the sinking has commonly been greater in the middle than near the walls. The walls themselves, whenever their foundations do not lie at a great depth, have been penetrated and undermined by worms, and have consequently subsided. The unequal subsidence thus caused, probably explains the great cracks which may be seen in many ancient walls, as well as their inclination from the perpendicular.

CHAPTER V.

THE ACTION OF WORMS IN THE DENUDATION OF THE LAND.

Evidence of the amount of denudation which the land has undergone—Subaerial denudation—The deposition of dust—Vegetable mould, its dark colour and fine texture largely due to the action of worms—The disintegration of rocks by the humus-acids—Similar acids apparently generated within the bodies of worms—The action of these acids facilitated by the continued movement of the particles of earth—A thick bed of mould checks the disintegration of the underlying soil and rocks. Particles of stone worn or triturated in the gizzards of worms—Swallowed stones serve as mill-stones—The levigated state of the castings—Fragments of brick in the castings over ancient buildings well rounded. The triturating power of worms not quite insignificant under a geological point of view.

No one doubts that our world at one time consisted of crystalline rocks, and that it is to their disintegration through the action of air, water, changes of temperature, rivers, waves of the sea, earthquakes and volcanic outbursts, that we owe our sedimentary formations. These after being consolidated and sometimes recrystallized, have often been again disintegrated. Denudation means the removal of such disintegrated matter to a lower level. Of the many striking results due to the modern progress of geology there are hardly any more striking than those which relate to denudation. It was long ago seen that there must have been an immense amount of denudation; but until the successive formations were carefully mapped and measured, no one fully realised how great was the amount. One of the first and most remarkable memoirs ever published on this subject was that by Ramsay,* who in 1846 showed that in Wales from 9000 to 11,000 feet in thickness of solid rock had been stripped off large tracks of country. Perhaps the plainest evidence of great denudation is afforded by faults or cracks, which extend for many miles across certain districts, with the strata on one side raised even ten thousand feet above the corresponding strata on the opposite side; and yet there is not a vestige of this gigantic displacement visible on the surface of the

* "On the denudation of South Wales," &c., 'Memoirs of the Geological Survey of Great Britain,' vol. i., p. 297, 1846.

land. A huge pile of rock has been planed away on one side and not a remnant left.

Until the last twenty or thirty years, most geologists thought that the waves of the sea were the chief agents in the work of denudation; but we may now feel sure that air and rain, aided by streams and rivers, are much more powerful agents,—that is if we consider the whole area of the land. The long lines of escarpment which stretch across several parts of England were formerly considered to be undoubtedly ancient coast-lines; but we now know that they stand up above the general surface merely from resisting air, rain and frost better than the adjoining formations. It has rarely been the good fortune of a geologist to bring conviction to the minds of his fellow-workers on a disputed point by a single memoir; but Mr. Whitaker, of the Geological Survey of England, was so fortunate when, in 1867, he published his paper "On sub-aerial Denudation, and on Cliffs and Escarpments of the Chalk."* Before this

* 'Geological Magazine,' October and November, 1867, vol. iv. pp. 447 and 483. Copious references on the subject are given in this remarkable memoir.

paper appeared, Mr. A. Tylor had adduced important evidence on sub-aerial denudation, by showing that the amount of matter brought down by rivers must infallibly lower the level of their drainage-basins by many feet in no immense lapse of time. This line of argument has since been followed up in the most interesting manner by Archibald Geikie, Croll and others, in a series of valuable memoirs.* For the sake of those who have never attended to this subject, a single instance may be here given, namely that of the Mississippi, which is chosen because the amount of sediment brought down by this great river has been investigated with especial care by order of the United States Government. The result is, as Mr. Croll shows, that the mean level of its enormous area of

* A. Tylor "On changes of the sea-level," &c., 'Philosophical Mag.' (Ser. 4th) vol. v., 1853, p. 258. Archibald Geikie, Transactions Geolog. Soc. of Glasgow, vol. iii., p. 153 (read March, 1868). Croll "On Geological Time," 'Philosophical Mag.', May, August, and November, 1868. See also Croll, 'Climate and Time,' 1875, Chap. XX. For some recent information on the amount of sediment brought down by rivers, see 'Nature,' Sept. 23rd, 1880. Mr. T. Mellard Reade has published some interesting articles on the astonishing amount of matter brought down in solution by rivers. See Address, Geolog. Soc., Liverpool, 1876–77.

drainage must be lowered $\frac{1}{4566}$ of a foot annually, or 1 foot in 4566 years. Consequently, taking the best estimate of the mean height of the North American continent, viz. 748 feet, and looking to the future, the whole of the great Mississippi basin will be washed away, and "brought down to the sea-"level in less than 4,500,000 years, if no "elevation of the land takes place." Some rivers carry down much more sediment relatively to their size, and some much less than the Mississippi.

Disintegrated matter is carried away by the wind as well as by running water. During volcanic outbursts much rock is triturated and is thus widely dispersed; and in all arid countries the wind plays an important part in the removal of such matter. Wind-driven sand also wears down the hardest rocks. I have shown * that during four months of the year a large quantity of dust is blown from the north-western shores of Africa, and falls on the Atlantic over a

* "An account of the fine dust which often falls on Vessels in the Atlantic Ocean," Proc. Geolog. Soc. of London, June 4th, 1845.

space of 1600 miles in latitude, and for a distance of from 300 to 600 miles from the coast. But dust has been seen to fall at a distance of 1030 miles from the shores of Africa. During a stay of three weeks at St. Jago in the Cape Verde Archipelago, the atmosphere was almost always hazy, and extremely fine dust coming from Africa was continually falling. In some of this dust which fell in the open ocean at a distance of between 330 and 380 miles from the African coast, there were many particles of stone, about $\frac{1}{1000}$ of an inch square. Nearer to the coast the water has been seen to be so much discoloured by the falling dust, that a sailing vessel left a track behind her. In countries, like the Cape Verde Archipelago, where it seldom rains and there are no frosts, the solid rock nevertheless disintegrates; and in conformity with the views lately advanced by a distinguished Belgian geologist, De Koninck, such disintegration may be attributed in chief part to the action of the carbonic and nitric acids, together with the nitrates and nitrites of ammonia, dissolved in the dew.

In all humid, even moderately humid,

countries, worms aid in the work of denudation in several ways. The vegetable mould which covers, as with a mantle, the surface of the land, has all passed many times through their bodies. Mould differs in appearance from the subsoil only in its dark colour, and in the absence of fragments or particles of stone (when such are present in the subsoil), larger than those which can pass through the alimentary canal of a worm. This sifting of the soil is aided, as has already been remarked, by burrowing animals of many kinds, especially by ants. In countries where the summer is long and dry, the mould in protected places must be largely increased by dust blown from other and more exposed places. For instance, the quantity of dust sometimes blown over the plains of La Plata, where there are no solid rocks, is so great, that during the "gran seco," 1827 to 1830, the appearance of the land, which is here unenclosed, was so completely changed that the inhabitants could not recognise the limits of their own estates, and endless lawsuits arose. Immense quantities of dust are likewise blown about in Egypt and in the south of France. In China, as Richthofen maintains, beds appearing like fine sediment, several hundred feet in thickness and extending over an enormous area, owe their origin to dust blown from the high lands of central Asia.* In humid countries like Great Britain, as long as the land remains in its natural state clothed with vegetation, the mould in any one place can hardly be much increased by dust; but in its present condition, the fields near high roads, where there is much traffic, must receive a considerable amount of dust, and when fields are harrowed during dry and windy weather, clouds of dust may be seen to be blown away. But in all these cases the surface-soil is merely transported from one place to another. The dust which falls so thickly within our houses con-

* For La Plata, see my 'Journal of Researches,' during the voyage of the *Beagle*, 1845, p. 133. Élie de Beaumont has given ('Leçons de Géolog. pratique,' tom. I. 1845, p. 183) an excellent account of the enormous quantity of dust which is transported in some countries. I cannot but think that Mr. Proctor has somewhat exaggerated ('Pleasant Ways in Science,' 1879, p. 379) the agency of dust in a humid country like Great Britain. James Geikie has given ('Prehistoric Europe,' 1880, p. 165) a full abstract of Richthofen's views, which, however, he disputes.

sists largely of organic matter, and if spread over the land would in time decay and disappear almost entirely. It appears, however, from recent observations on the snow-fields of the Arctic regions, that some little meteoric dust of extra mundane origin is continually falling.

The dark colour of ordinary mould is obviously due to the presence of decaying organic matter, which, however, is present in but small quantities. The loss of weight which mould suffers when heated to redness seems to be in large part due to water in combination being dispelled. In one sample of fertile mould the amount of organic matter was ascertained to be only 1·76 per cent.; in some artificially prepared soil it was as much as 5·5 per cent., and in the famous black soil of Russia from 5 to even 12 per cent.* In leaf-mould formed exclusively by the decay of leaves the amount is much greater, and in peat the carbon alone sometimes amounts to

* These statements are taken from Von Hensen in 'Zeitschrift für wissenschaft. Zoologie,' Bd. xxviii., 1877, p. 360. Those with respect to peat are taken from Mr. A. A. Julien in 'Proc. American Assoc. Science,' 1879, p. 314.

64 per cent.; but with these latter cases we are not here concerned. The carbon in the soil tends gradually to oxidise and to disappear, except where water accumulates and the climate is cool;* so that in the oldest pasture-land there is no great excess of organic matter, notwithstanding the continued decay of the roots and the underground stems of plants, and the occasional addition of manure. The disappearance of the organic matter from mould is probably much aided by its being brought again and again to the surface in the castings of worms.

Worms, on the other hand, add largely to the organic matter in the soil by the astonishing number of half-decayed leaves which they draw into their burrows to a depth of 2 or 3 inches. They do this chiefly for obtaining food, but partly for closing the mouths of their burrows and for lining the upper part. The leaves which they consume are moistened, torn into small shreds, partially digested, and intimately commingled with

* I have given some facts on the climate necessary or favourable for the formation of peat, in my 'Journal of Researches,' 1845, p. 287.

earth; and it is this process which gives to vegetable mould its uniform dark tint. It is known that various kinds of acids are generated by the decay of vegetable matter; and from the contents of the intestines of worms and from their castings being acid, it seems probable that the process of digestion induces an analogous chemical change in the swallowed, triturated, and half decayed leaves. The large quantity of carbonate of lime secreted by the calciferous glands apparently serves to neutralise the acids thus generated; for the digestive fluid of worms will not act unless it be alkaline. As the contents of the upper part of their intestines are acid, the acidity can hardly be due to the presence of uric acid. We may therefore conclude that the acids in the alimentary canal of worms are formed during the digestive process; and that probably they are nearly of the same nature as those in ordinary humus. The latter are well known to have the power of de-oxidising or dissolving peroxide of iron, as may be seen wherever peat overlies red sand, or where a rotten root penetrates such sand. Now I kept some worms in a pot filled with very fine reddish sand, consisting of minute particles of silex coated with the red oxide of iron; and the burrows, which the worms made through this sand, were lined or coated in the usual manner with their castings, formed of the sand mingled with their intestinal secretions and the refuse of the digested leaves; and this sand had almost wholly lost its red colour. When small portions of it were placed under the microscope, most of the grains were seen to be transparent and colourless, owing to the dissolution of the oxide; whilst almost all the grains taken from other parts of the pot were coated with the oxide. Acetic acid produced hardly any effect on this sand; and even hydrochloric, nitric and sulphuric acids, diluted as in the Pharmacopœia, produced less effect than did the acids in the intestines of the worms.

Mr. A. A. Julien has lately collected all the extant information about the acids generated in humus, which, according to some chemists, amount to more than a dozen different kinds. These acids, as well as their acid salts (i.e., in combination with potash, soda, and ammonia), act energetically on carbonate of lime and on the oxides of iron. It is, also, known that some of these acids, which were called long ago by Thénard azo-humic, are enabled to dissolve colloid silica in proportion to the nitrogen which they contain.* In the formation of these latter acids worms probably afford some aid, for Dr. H. Johnson informs me that by Nessler's test he found 0·018 per cent. of ammonia in their castings.

The several humus-acids, which appear, as we have just seen, to be generated within the bodies of worms during the digestive process, and their acid salts, play a highly important part, according to the recent observations of Mr. Julien, in the disintegration of various kinds of rocks. It has long been known that the carbonic acid, and no doubt nitric and nitrous acids, which are present in rain-water, act in like manner. There is, also, a great excess of carbonic acid in all soils, especially in rich soils, and this is dissolved by the water

* A. A. Julien "On the Geological action of the Humus-acids," 'Proc. American Assoc. Science,' vol. xxviii., 1879, p. 311. Also on "Chemical erosion on Mountain Summits;" 'New York Academy of Sciences,' Oct. 14, 1878, as quoted in the 'American Naturalist.' See also, on this subject, S. W. Johnson, "How Crops Feed," 1870, p. 138.

in the ground. The living roots of plants, moreover, as Sachs and others have shown, quickly corrode and leave their impressions on polished slabs of marble, dolomite and phosphate of lime. They will attack even basalt and sandstone.* But we are not here concerned with agencies which are wholly independent of the action of worms.

The combination of any acid with a base is much facilitated by agitation, as fresh surfaces are thus continually brought into contact. This will be thoroughly effected with the particles of stone and earth in the intestines of worms, during the digestive process; and it should be remembered that the entire mass of the mould over every field, passes, in the course of a few years, through their alimentary canals. Moreover as the old burrows slowly collapse, and as fresh castings are continually brought to the surface, the whole superficial layer of mould slowly revolves or circulates; and the friction of the particles one with another will rub off the finest films of disintegrated matter as soon as

* See, for references on this subject, S. W. Johnson, "How Crops Feed," 1870, p. 326.

they are formed. Through these several means, minute fragments of rocks of many kinds and mere particles in the soil will be continually exposed to chemical decomposition; and thus the amount of soil will tend to increase.

As worms line their burrows with their castings, and as the burrows penetrate to a depth of 5 or 6, or even more feet, some small amount of the humus-acids will be carried far down, and will there act on the underlying rocks and fragments of rock. Thus the thickness of the soil, if none be removed from the surface, will steadily though slowly tend to increase; but the accumulation will after a time delay the disintegration of the underlying rocks and of the more deeply seated particles. For the humus-acids which are generated chiefly in the upper layer of vegetable mould, are extremely unstable compounds, and are liable to decomposition before they reach any considerable depth.* A thick bed of overlying soil will also check the downward extension of great fluctuations of temperature, and in cold countries will check

* This statement is taken from Mr. Julien, 'Proc. American Assoc. Science,' vol. xxviii., 1879, p. 330.

the powerful action of frost. The free access of air will likewise be excluded. From these several causes disintegration would be almost arrested, if the overlying mould were to increase much in thickness, owing to none or little being removed from the surface.* In my own immediate neighbourhood we have a curious proof how effectually a few feet of clay checks some change which goes on in flints, lying freely exposed; for the large ones which have lain for some time on the surface of ploughed fields cannot be used for building; they will not cleave properly and are said by the workmen to be rotten.† It is

* The preservative power of a layer of mould and turf is often shown by the perfect state of the glacial scratches on rocks when first uncovered. Mr. J. Geikie maintains, in his last very interesting work ('Prehistoric Europe,' 1881), that the more perfect scratches are probably due to the last access of cold and increase of ice, during the long-continued, intermittent glacial period.

† Many geologists have felt much surprise at the complete disappearance of flints over wide and nearly level areas, from which the chalk has been removed by subaerial denudation. But the surface of every flint is coated by an opaque modified layer, which will just yield to a steel point, whilst the freshly-fractured, translucent surface will not thus yield. The removal by atmospheric agencies of the outer modified surfaces of freely exposed flints, though no doubt excessively slow, together with the modification travelling inwards, will, as may be suspected, ultimately lead to their complete disintegration, notwithstanding that they appear to be so extremely durable.

therefore necessary to obtain flints for building purposes from the bed of red clay overlying the chalk (the residue of its dissolution by rainwater) or from the chalk itself.

Not only do worms aid indirectly in the chemical disintegration of rocks, but there is good reason to believe that they likewise act in a direct and mechanical manner on the smaller particles. All the species which swallow earth are furnished with gizzards; and these are lined with so thick a chitinous membrane, that Perrier speaks of it,* as "une véritable armature." The gizzard is surrounded by powerful transverse muscles, which, according to Claparède, are about ten times as thick as the longitudinal ones; and Perrier saw them contracting energetically. Worms belonging to one genus, Digaster, have two distinct but quite similar gizzards; and in another genus, Moniligaster, the second gizzard consists of four pouches, one succeeding the other, so that it may almost be said to have five gizzards.† In the same

* 'Archives de Zoolog. expér.' tom. iii. 1874, p. 409.

† 'Nouvelles Archives du Muséum,' tom. viii. 1872, p. 95, 131.

manner as gallinaceous and struthious birds swallow stones to aid in the trituration of their food, so it appears to be with terricolous worms. The gizzards of thirty-eight of our common worms were opened, and in twenty-five of them small stones or grains of sand, sometimes together with the hard calcareous concretions formed within the anterior calciferous glands, were found, and in two others concretions alone. In the gizzards of the remaining worms there were no stones; but some of these were not real exceptions, as the gizzards were opened late in the autumn, when the worms had ceased to feed and their gizzards were quite empty.*

When worms make their burrows through earth abounding with little stones, no doubt many will be unavoidably swallowed; but it must not be supposed that this fact accounts for the frequency with which stones and sand are found in their gizzards. For beads of glass and fragments of brick and of hard tiles were scattered over the surface

* Morren, in speaking of the earth in the alimentary canals of worms, says, "præsepè cum lapillis commixtam vidi:" 'De Lumbrici terrestris,' &c., 1829, p. 16.

of the earth, in pots in which worms were kept and had already made their burrows; and very many of these beads and fragments were picked up and swallowed by the worms, for they were found in their castings, intestines, and gizzards. They even swallowed the coarse red dust, formed by the pounding of the tiles. Nor can it be supposed that they mistook the beads and fragments for food; for we have seen that their taste is delicate enough to distinguish between different kinds of leaves. It is therefore manifest that they swallow hard objects, such as bits of stone, beads of glass and angular fragments of bricks or tiles for some special purpose; and it can hardly be doubted that this is to aid their gizzards in crushing and grinding the earth, which they so largely consume. That such hard objects are not necessary for crushing leaves, may be inferred from the fact that certain species, which live in mud or water and feed on dead or living vegetable matter, but which do not swallow earth, are not provided with gizzards,* and

* Perrier, 'Archives de Zoolog. expér.' tom. iii. 1874, p. 419.

therefore cannot have the power of utilising stones.

During the grinding process, the particles of earth must be rubbed against one another, and between the stones and the tough lining membrane of the gizzard. The softer particles will thus suffer some attrition, and will perhaps even be crushed. This conclusion is supported by the appearance of freshly ejected castings, for these often reminded me of the appearance of paint which has just been ground by a workman between two flat stones. Morren remarks that the intestinal canal is "impleta tenuissimâ terrâ, veluti in pulverem redactâ." * Perrier also speaks of "l'état de pâte excessivement fine à laquelle est réduite la terre qu'ils rejettent," &c.†

As the amount of trituration which the particles of earth undergo in the gizzards of worms possesses some interest (as we shall hereafter see), I endeavoured to obtain evidence on this head by carefully examining many of the fragments which had passed

* Morren, 'De Lumbrici terrestris,' &c., p. 16.

† 'Archives de Zoolog. Expér.' tom. iii. 1874, p. 418.

through their alimentary canals. With worms living in a state of nature, it is of course impossible to know how much the fragments may have been worn before they were swallowed. It is, however, clear that worms do not habitually select already rounded particles, for sharply angular bits of flint and of other hard rocks were often found in their gizzards or intestines. On three occasions sharp spines from the stems of rose-bushes were thus found. Worms kept in confinement repeatedly swallowed angular fragments of hard tile, coal, cinders, and even the sharpest fragments of glass. Gallinaceous and struthious birds retain the same stones in their gizzards for a long time, which thus become well rounded; but this does not appear to be the case with worms, judging from the large number of the fragments of tiles, glass beads, stones, &c., commonly found in their castings and intestines. So that unless the same fragments were to pass repeatedly through their gizzards, visible signs of attrition in the fragments could hardly be expected, except perhaps in the case of very soft stones.

I will now give such evidence of attrition as I have been able to collect. In the gizzards of some worms dug out of a thin bed of mould over the chalk, there were many well-rounded small fragments of chalk, and two fragments of the shells of a land-mollusc (as ascertained by their microscopical structure), which latter were not only rounded but somewhat polished. The calcareous concretions formed in the calciferous glands, which are often found in their gizzards, intestines, and occasionally in their castings, when of large size, sometimes appeared to have been rounded; but with all calcareous bodies the rounded appearance may be partly or wholly due to their corrosion by carbonic acid and the humus-acids. In the gizzards of several worms collected in my kitchen garden near a hothouse, eight little fragments of cinders were found, and of these, six appeared more or less rounded, as were two bits of brick; but some other bits were not at all rounded. A farm-road near Abinger Hall had been covered seven years before with brick-rubbish to the depth of about 6 inches; turf had grown over this rubbish on both sides of the road for a width of 18 inches, and on this turf there were innumerable castings. Some of them were coloured of a uniform red owing to the presence of much brick-dust, and they contained many particles of brick and of hard mortar from 1 to 3 mm. in diameter, most of which were plainly rounded; but all these particles may have been rounded before they were protected by the turf and were swallowed, like those on the bare parts of the road which were much worn. A hole in a pasture-field had been filled up with brick-rubbish at the same time, viz., seven years ago, and was now covered with turf; and here the castings contained very many particles of brick, all more or less rounded; and this brick-rubbish, after being shot into the hole, could not have undergone any attrition. Again, old bricks very little broken, together with fragments of mortar, were laid down to form walks, and were then covered with from 4 to 6 inches of gravel; six little fragments of brick were extracted from castings collected on these walks, three of which were plainly worn.

There were also very many particles of hard mortar, about half of which were well rounded; and it is not credible that these could have suffered so much corrosion from the action of carbonic acid in the course of only seven years.

Much better evidence of the attrition of hard objects in the gizzards of worms, is afforded by the state of the small fragments of tiles or bricks, and of concrete in the castings thrown up where ancient buildings once stood. As all the mould covering a field passes every few years through the bodies of worms, the same small fragments will probably be swallowed and brought to the surface many times in the course of centuries. It should be premised that in the several following cases, the finer matter was first washed away from the castings, and then *all* the particles of bricks, tiles and concrete were collected without any selection, and were afterwards examined. Now in the castings ejected between the tesseræ on one of the buried floors of the Roman villa at Abinger, there were many particles (from ½ to 2 mm. in diameter) of tiles and concrete, which it was impossible to look at with the naked eye or through a strong lens, and doubt for a moment that they had almost all undergone much attrition. I speak thus after having examined small water-worn pebbles, formed from Roman bricks, which M. Henri de Saussure had the kindness to send me, and which he had extracted from sand and gravel beds, deposited on the shores of the Lake of Geneva, at a former period when the water stood at about two metres above its present level. The smallest of these water-worn pebbles of brick from Geneva resembled closely many of those extracted from the gizzards of worms, but the larger ones were somewhat smoother.

Four castings found on the recently uncovered, tesselated floor of the great room in the Roman villa at Brading, contained many particles of tile or brick, of mortar, and of hard white cement; and the majority of these appeared plainly worn. The particles of mortar, however, seemed to have suffered more corrosion than attrition, for grains of silex often projected from their surfaces. Castings from within the nave of Beaulieu

Abbey, which was destroyed by Henry VIII., were collected from a level expanse of turf, overlying the buried tesselated pavement, through which worm-burrows passed; and these castings contained innumerable particles of tiles and bricks, of concrete and cement, the majority of which had manifestly undergone some or much attrition. There were also many minute flakes of a micaceous slate, the points of which were rounded. If the above supposition, that in all these cases the same minute fragments have passed several times through the gizzards of worms, be rejected, notwithstanding its inherent probability, we must then assume that in all the above cases the many rounded fragments found in the castings had all accidentally undergone much attrition before they were swallowed; and this is highly improbable.

On the other hand it must be stated that fragments of ornamental tiles, somewhat harder than common tiles or bricks, which had been swallowed only once by worms kept in confinement, were with the doubtful exception of one or two of the smallest grains, not at all rounded. Nevertheless some of them appeared a little worn, though not rounded. Notwithstanding these cases, if we consider the evidence above given, there can be little doubt that the fragments, which serve as millstones in the gizzards of worms, suffer, when of a not very hard texture, some amount of attrition; and that the smaller particles in the earth, which is habitually swallowed in such astonishingly large quantities by worms, are ground together and are thus levigated. If this be the case, the "terra tenuissima,"—the "pâte excessivement fine,"—of which the castings largely consist, is in part due to the mechanical action of the gizzard; * and this fine matter, as we shall see in the next chapter, is that which is chiefly washed away from the innumerable castings on every field during each heavy shower of rain. If the softer stones yield at all, the harder ones will suffer some slight amount of wear and tear.

* This conclusion reminds me of the vast amount of extremely fine chalky mud which is found within the lagoons of many atolls, where the sea is tranquil and waves cannot triturate the blocks of coral. This mud must, as I believe ('The Structure and Distribution of Coral-Reefs,' 2nd edit. 1874, p. 19), be attributed to the innumerable annelids and other animals which burrow into the dead coral, and to the fishes, Holothurians, &c., which browse on the living corals.

The trituration of small particles of stone in the gizzards of worms is of more importance under a geological point of view than may at first appear to be the case; for Mr. Sorby has clearly shown that the ordinary means of disintegration, namely running water and the waves of the sea, act with less and less power on fragments of rock the smaller they are. "Hence," as he remarks, "even making no allowance for the extra "buoying up of very minute particles by a "current of water, depending on surface "cohesion, the effects of wearing on the form "of the grains must vary directly as their "diameter or thereabouts. If so, a grain $\frac{1}{10}$ "of an inch in diameter would be worn ten "times as much as one $\frac{1}{100}$ of an inch in "diameter, and at least a hundred times as "much as one $\frac{1}{1000}$ of an inch in diameter. "Perhaps, then, we may conclude that a "grain $\frac{1}{10}$ of an inch in diameter would be "worn as much or more in drifting a mile as "a grain $\frac{1}{1000}$ of an inch in being drifted "100 miles. On the same principle a pebble "one inch in diameter would be worn relatively more by being drifted only a few "hundred yards."* Nor should we forget, in considering the power which worms exert in triturating particles of rock, that there is good evidence that on each acre of land, which is sufficiently damp and not too sandy, gravelly or rocky for worms to inhabit, a weight of more than ten tons of earth annually passes through their bodies and is brought to the surface. The result for a country of the size of Great Britain, within a period not very long in a geological sense, such as a million years, cannot be insignificant; for the ten tons of earth has to be multiplied first by the above number of years, and then by the number of acres fully stocked with worms; and in England, together with Scotland, the land which is cultivated and is well fitted for these animals, has been estimated at above 32 million acres. The product is 320 million million tons of earth.

* Anniversary Address: 'The Quarterly Journal of the Geological Soc.' May 1880, p. 59.

CHAPTER VI.

THE DENUDATION OF THE LAND—*continued*.

Denudation aided by recently ejected castings flowing down inclined grass-covered surfaces—The amount of earth which annually flows downwards—The effect of tropical rain on worm castings—The finest particles of earth washed completely away from castings—The disintegration of dried castings into pellets, and their rolling down inclined surfaces—The formation of little ledges on hill-sides, in part due to the accumulation of disintegrated castings—Castings blown to leeward over level land—An attempt to estimate the amount thus blown—The degradation of ancient encampments and tumuli—The preservation of the crowns and furrows on land anciently ploughed—The formation and amount of mould over the Chalk formation.

We are now prepared to consider the more direct part which worms take in the denudation of the land. When reflecting on subaerial denudation, it formerly appeared to me, as it has to others, that a nearly level or very gently inclined surface, covered with turf, could suffer no loss during even a long lapse of time. It may, however, be urged that at long intervals, debacles of rain or water-spouts would remove all the mould from a very gentle slope; but when examining the steep, turf-covered slopes in Glen Roy, I was struck with the fact how rarely any such event could have happened since the Glacial period, as was plain from the well-preserved state of the three successive "roads" or lake-margins. But the difficulty in believing that earth in any appreciable quantity can be removed from a gently inclined surface, covered with vegetation and matted with roots, is removed through the agency of worms. For the many castings which are thrown up during rain, and those thrown up some little time before heavy rain, flow for a short distance down an inclined surface. Moreover much of the finest levigated earth is washed completely away from the castings. During dry weather castings often disintegrate into small rounded pellets, and these from their weight often roll down any slope. This is more especially apt to occur when they are started by the wind, and probably when started by the touch of an animal, however small. We shall also see that a strong wind blows all the castings, even on a level field, to leeward, whilst they are soft; and in like manner the pellets when they are dry. If the wind blows in nearly the direction of an inclined surface, the flowing down of the castings is much aided.

The observations on which these several statements are founded must now be given in some detail. Castings when first ejected are viscid and soft; during rain, at which time worms apparently prefer to eject them, they are still softer; so that I have sometimes thought that worms must swallow much water at such times. However this may be, rain, even when not very heavy, if long continued, renders recently-ejected castings semi-fluid; and on level ground they then spread out into thin, circular, flat discs, exactly as would so much honey or very soft mortar, with all traces of their vermiform structure lost. This latter fact was sometimes made evident, when a worm had subsequently bored through a flat circular disc of this kind, and heaped up a fresh vermiform mass in the centre. These flat subsided discs have been repeatedly seen by me after heavy rain, in many places on land of all kinds.

On the flowing of wet castings, and the rolling of dry disintegrated castings down inclined surfaces.—When castings are ejected on an inclined surface during or shortly before heavy rain, they cannot fail to flow a little down the slope. Thus, on some steep slopes in Knowle Park, which were covered with coarse grass and had apparently existed in this state from time immemorial, I found (Oct. 22, 1872) after several wet days that almost all the many castings were considerably elongated in the line of the slope; and that they now consisted of smooth, only slightly conical masses. Whenever the mouths of the burrows could be found from which the earth had been ejected, there was more earth below than above them. After some heavy storms of rain (Jan. 25, 1872) two rather steeply inclined fields near Down, which had formerly been ploughed and were now rather sparsely clothed with poor grass, were visited, and many castings extended down the slopes for a length of 5 inches, which was twice or thrice the usual diameter of the castings thrown up on the level parts of these same fields. On some fine grassy slopes in Holwood Park, inclined at angles between 8° and 11° 30′ with the horizon, where the surface apparently had never been disturbed by the hand of man, castings abounded in extraordinary numbers: and a space 16 inches in length transversely to the slope and 6 inches in the line of the slope, was completely coated, between the blades of grass, with a uniform sheet of confluent and subsided castings. Here also in many places the castings had flowed down the slope, and now formed smooth narrow patches of earth, 6, 7, and 7½ inches in length. Some of these consisted of two castings, one above the other, which had become so completely confluent that they could hardly be distinguished. On my lawn, clothed with very fine grass, most of the castings are black, but some are yellowish from earth having been brought up from a greater depth than usual, and the flowing-down of these yellow castings after heavy rain, could be clearly seen where the slope was 5°; and where it was less than 1° some evidence of their flowing down could still be detected. On another occasion, after rain which was never heavy, but which lasted for 18 hours, all the castings on this same gently inclined lawn had lost their vermiform structure; and they had flowed, so that fully two-thirds of the ejected earth lay below the mouths of the burrows.

These observations led me to make others with more care. Eight castings were found on my lawn, where the grass-blades are fine and close together, and three others on a field with coarse grass. The inclination of the surface at the eleven places where these castings were collected varied between 4° 30′ and 17° 30′; the mean of the eleven inclinations being 9° 26′. The length of the castings in the direction of the slope was first measured with as much accuracy as their irregularities would permit. It was found possible to make these measurements within about $\frac{1}{8}$ of an inch, but one of the castings was too irregular to admit of measurement. The average length in the direction of the slope of the remaining ten castings was 2·03 inches. The castings were then divided with a knife into two parts along a horizontal line passing through the mouth of the burrow, which was discovered by slicing off the turf; and all the ejected earth was separately collected, namely the part above the hole and the part below. Afterwards these two parts were weighed. In every case there was much more earth below than above; the mean weight of that above being 103 grains, and of that below 205 grains; so that the latter was very nearly double the former. As on level ground castings are commonly thrown up almost equally round the mouths of the burrows, this difference in weight indicates the amount of ejected earth which had flowed down the slope. But very many more observations would be requisite to arrive at any general result; for the nature of the vegetation and other accidental circumstances, such as the heaviness of the rain, the direction and force of the wind, &c., appear to be more important in determining the quantity of the earth which flows down a slope than its angle. Thus with four castings on my lawn (included in the above eleven) where the mean slope was 7° 19′, the difference in the amount of earth above and below the burrows was greater than with three other castings on the same lawn where the mean slope was 12° 5′.

We may, however, take the above eleven cases, which are accurate as far as they go, and calculate the weight of the ejected earth which annually flows down a slope having a mean inclination of 9° 26′. This was done by my son George. It has been shown that almost exactly two-thirds of the ejected earth is found below the mouth of the burrow and one-third above it. Now if the two-thirds which is below the hole be divided into two equal parts, the upper half of this two-thirds exactly counterbalances the one-third which is above the hole, so that as far as regards the one-third above and the upper half of the two-thirds below, there is no flow of earth down the hill-side. The earth constituting the lower half of the two-thirds is, however, displaced through distances which are different for every part of it, but which may be represented by the distance between the middle point of the lower half of the two-thirds and the hole. So that the average distance of displacement is a half of the whole length of the worm-casting. Now the average length of ten out of the above eleven castings was 2·03 inches, and half of this we may take as being one inch. It may therefore be concluded that one-third of the whole earth brought to the surface was in these cases carried down the slope through one inch.

It was shown in the third chapter that on Leith Hill Common, dry earth weighing at least 7·453 lbs. was brought up by worms to the surface on a square yard in the course of a year. If a square yard be drawn on a hill-side with two of its sides horizontal, then it is clear that only $\frac{1}{36}$ part of the earth brought up on that square yard would be near enough to its lower side to cross it, supposing the displacement of the earth to be through one inch. But it appears that only $\frac{1}{3}$ of the earth brought up can be considered to flow downwards; hence $\frac{1}{3}$ of $\frac{1}{36}$ or $\frac{1}{108}$ of 7·453 lbs. will cross the lower side of our square yard in a year. Now $\frac{1}{108}$ of 7·453 lbs. is 1·1 oz. Therefore 1·1 oz. of dry earth will annually cross each linear yard running horizontally along a slope having the above inclination; or very nearly 7 lbs. will annually cross a horizontal line, 100 yards in length, on a hill-side having this inclination.

A more accurate, though still very rough, calculation can be made of the bulk of earth, which in its natural damp state annually flows down the same slope over a yard-line drawn horizontally across it. From the several cases given in the third chapter, it is known that the castings annually brought to the surface on a square yard, if uniformly spread out would form a layer ·2 of an inch in thickness: it therefore follows by a calculation similar to the one already given, that $\frac{1}{3}$ of $\cdot 2 \times 36$, or 2·4 cubic inches of damp earth will annually cross a horizontal line one yard in length on a hill-side with the above inclination. This bulk of damp castings was found to weigh 1·85 oz. Therefore 11·56 lbs. of damp earth, instead of 7 lbs. of dry earth as by the former calculation, would annually cross a line 100 yards in length on our inclined surface.

In these calculations it has been assumed that the castings flow a short distance downwards during the whole year, but this occurs only with those ejected during or shortly before rain; so that the above results are thus far exaggerated. On the other hand, during rain much of the finest earth is washed to a considerable distance from the castings, even where the slope is an extremely gentle one, and is thus wholly lost as far as the above calculations are concerned. Castings ejected during dry weather and which have set hard, lose in the same manner a considerable quantity of fine earth. Dried castings, moreover, are apt to disintegrate into little pellets, which often roll or are blown down any inclined surface. Therefore the above result, namely that 2·4 cubic inches of earth (weighing 1·85 oz. whilst damp) annually crosses a yard-line of the specified kind, is probably not much if at all exaggerated.

This amount is small; but we should bear in mind how many branching valleys intersect most countries, the whole length of which must be very great; and that earth is steadily travelling down both turf-covered sides of each valley. For every 100 yards in length in a valley with sides sloping as in the foregoing cases, 480 cubic inches of damp earth, weighing above 23 pounds, will annually reach the bottom. Here a thick bed of alluvium will accumulate, ready to be washed away in the course of centuries, as the stream in the middle meanders from side to side.

If it could be shown that worms generally excavate their burrows at right angles to an inclined surface, and this would be their shortest course for bringing up earth from beneath, then as the old burrows collapsed from the weight of the superincumbent soil, the collapsing would inevitably cause the whole bed of vegetable mould to sink or slide slowly down the inclined surface. But to ascertain the direction of many burrows was found too difficult and troublesome. A straight piece of wire was, however, pushed into twenty-five burrows on several sloping fields, and in eight cases the burrows were nearly at right angles to the slope; whilst in the remaining cases they were indifferently directed at various angles, either upwards or downwards with respect to the slope.

In countries where the rain is very heavy,

as in the tropics, the castings appear, as might have been expected, to be washed down in a greater degree than in England. Mr. Scott informs me that near Calcutta the tall columnar castings (previously described). the diameter of which is usually between 1 and 1½ inch, subside on a level surface, after heavy rain, into almost circular, thin, flat discs, between 3 and 4 and sometimes 5 inches in diameter. Three fresh castings, which had been ejected in the Botanic Gardens "on a slightly inclined, grass-"covered, artificial bank of loamy clay," were carefully measured, and had a mean height of 2·17, and a mean diameter of 1·43 inches; these after heavy rain, formed elongated patches of earth, with a mean length in the direction of the slope of 5·83 inches. As the earth had spread very little up the slope, a large part, judging from the original diameter of these castings, must have flowed bodily downwards about 4 inches. Moreover some of the finest earth of which they were composed must have been washed completely away to a still greater distance. In drier sites near Calcutta, a species of worm ejects its castings, not in vermiform masses, but in little pellets of varying sizes: these are very numerous in some places, and Mr. Scott says that they "are washed away by every "shower."

I was led to believe that a considerable quantity of fine earth is washed quite away from castings during rain, from the surfaces of old ones being often studded with coarse particles. Accordingly a little fine precipitated chalk, moistened with saliva or gum-water, so as to be slightly viscid and of the same consistence as a fresh casting, was placed on the summits of several castings and gently mixed with them. These castings were then watered through a very fine rose, the drops from which were closer together than those of rain, but not nearly so large as those in a thunder storm; nor did they strike the ground with nearly so much force as drops during heavy rain. A casting thus treated subsided with surprising slowness, owing as I suppose to its viscidity. It did not flow bodily down the grass-covered surface of the lawn, which was here inclined at an angle of 16° 20′; nevertheless many particles of the chalk were found three inches below the casting. The experiment was repeated on three other castings on different parts of the lawn, which sloped at 2° 30′, 3° and 6°; and particles of chalk could be seen between 4 and 5 inches below the casting; and after the surface had become dry, particles were found in two cases at a distance of 5 and 6 inches. Several other castings with precipitated chalk placed on their summits were left to the natural action of the rain. In one case, after rain which was not heavy, the casting was longitudinally streaked with white. In two other cases the surface of the ground was rendered somewhat white for a distance of one inch from the casting; and some soil collected at a distance of 2½ inches, where the slope was 7°, effervesced slightly when placed in acid. After one or two weeks, the chalk was wholly or almost wholly washed away from all the castings on which it had been placed, and these had recovered their natural colour.

It may be here remarked that after very heavy rain shallow pools may be seen on level or nearly level fields, where the soil is not very porous, and the water in them is often slightly muddy; when such little pools have dried, the leaves and blades of grass at their bottoms are generally coated with a thin layer of mud. This mud I believe is derived in large part from recently ejected castings.

Dr. King informs me that the majority of the before described gigantic castings, which he found on a fully exposed, bare, gravelly knoll on the Nilgiri Mountains in India, had been more or less weathered by the previous north-east monsoon; and most of them presented a subsided appearance. The worms here eject their castings only during the rainy season; and at the time of Dr. King's visit no rain had fallen for 110 days. He carefully examined the ground between the place where these huge castings lay, and a little water-course at the base of the knoll, and nowhere was there any accumulation of fine earth, such as would necessarily have been left by the disintegration of the castings if they had not been wholly removed. He therefore has no hesitation in asserting that the whole of these huge castings are annually washed during the two monsoons (when about 100 inches of rain fall) into the little water-course, and thence into the plains lying below at a depth of 3000 or 4000 feet.

Castings ejected before or during dry weather become hard, sometimes surprisingly hard, from the particles of earth having been cemented together by the intestinal secretions. Frost seems to be less effective in their disintegration than might have been expected. Nevertheless they readily disintegrate into small pellets, after being alternately moistened with rain and again dried. Those which have flowed during rain down a slope, disintegrate in the same manner. Such pellets often roll a little down any sloping surface; their descent being sometimes much aided by the wind. The whole bottom of a broad dry ditch in my grounds, where there were very few fresh castings, was completely covered with these pellets or disintegrated castings, which had rolled down the steep sides, inclined at an angle of 27°.

Near Nice, in places where the great cylindrical castings, previously described, abound, the soil consists of very fine arenaceo-calcareous loam; and Dr. King informs me that these castings are extremely liable to crumble during dry weather into small fragments, which are soon acted on by rain, and then sink down so as to be no longer distinguishable from the surrounding soil. He sent me a mass of such disintegrated castings, collected on the top of a bank, where none could have rolled down from above. They must have been ejected within the previous five or six months, but they now consisted of more or less rounded fragments of all sizes, from $\frac{3}{4}$ of an inch in diameter to minute grains and mere dust. Dr. King witnessed the crumbling process whilst drying some perfect castings, which he afterwards sent me. Mr. Scott also remarks on the crumbling of the castings near Calcutta and on the mountains of Sikkim during the hot and dry season.

When the castings near Nice had been ejected on an inclined surface, the disintegrated fragments rolled downwards, without losing their distinctive shape; and in some places could "be collected in basketfuls." Dr. King observed a striking instance of this fact on the Corniche road, where a drain, about $2\frac{1}{2}$ feet wide and 9 inches deep, had been made to catch the surface drainage from the adjoining hill-side. The bottom of this drain was covered for a distance of several hundred yards, to a depth of from 1½ to 3 inches, by a layer of broken castings, still retaining their characteristic shape. Nearly all these innumerable fragments had rolled down from above, for extremely few castings had been ejected in the drain itself. The hill-side was steep, but varied much in inclination, which Dr. King estimated at from 30° to 60° with the horizon. He climbed up the slope, and "found every here and there little embank-"ments, formed by fragments of the castings "that had been arrested in their downward "progress by irregularities of the surface, "by stones, twigs, &c. One little group of "plants of *Anemone hortensis* had acted in this "manner, and quite a small bank of soil had "collected round it. Much of this soil had "crumbled down, but a great deal of it still "retained the form of castings." Dr. King dug up this plant, and was struck with the thickness of the soil which must have recently accumulated over the crown of the rhizoma, as shown by the length of the bleached petioles, in comparison with those of other plants of the same kind, where there had been no such accumulation. The earth thus accumulated had no doubt been secured (as I have everywhere seen) by the smaller roots of the plants. After describing this and other analogous cases, Dr. King concludes: "I can have no doubt that worms "help greatly in the process of denudation."

Ledges of earth on steep hill-sides.—Little horizontal ledges, one above another, have been observed on steep grassy slopes in many parts of the world. Their formation has been attributed to animals travelling repeatedly along the slope in the same horizontal lines while grazing, and that they do thus move and use the ledges is certain; but Professor Henslow (a most careful observer) told Sir J. Hooker that he was convinced that this was not the sole cause of their formation. Sir J. Hooker saw such ledges on the Himalayan and Atlas ranges, where there were no domesticated animals and not many wild ones; but these latter would, it is probable, use the ledges at night while grazing like our domesticated animals. A friend observed for me the ledges on the Alps of Switzerland, and states that they ran at 3 or 4 ft. one above the other, and were about a foot in breadth. They had been deeply pitted by the feet of grazing cows. Similar ledges were observed by the same friend on our Chalk downs, and on an old talus of chalk-fragments (thrown out of a quarry) which had become clothed with turf.

My son Francis examined a Chalk escarpment near Lewes; and here on a part which was very steep, sloping at 40° with the horizon, about 30 flat ledges extended horizontally for more than 100 yards, at an average distance of about 20 inches, one beneath the other. They were from 9 to 10 inches in breadth. When viewed from a distance they presented a striking appearance, owing to their parallelism; but when examined closely, they were seen to be somewhat sinuous, and one often ran into another, giving the appearance of one ledge having forked into two. They are formed of light-coloured earth, which on the outside, where thickest, was in one case 9 inches, and in another case between 6 and 7 inches in thickness. Above the ledges, the thickness of the earth over the chalk was in the former case 4 and in the latter only 3 inches. The grass grew more vigorously on the outer edges of the ledges than on any other part of the slope, and here formed a tufted fringe. Their middle part was bare, but whether this had been caused by the trampling of sheep, which sometimes frequent the ledges, my son could not ascertain. Nor could he feel sure how much of the earth on the middle and bare parts, consisted of disintegrated worm-castings which had rolled down from above; but he felt convinced that some had thus originated; and it was manifest that the ledges with their grass-fringed edges would arrest any small object rolling down from above.

At one end or side of the bank bearing these ledges, the surface consisted in parts of bare chalk, and here the ledges were very irregular. At the other end of the bank, the slope suddenly became less steep, and here the ledges ceased rather abruptly; but little embankments only a foot or two in length were still present. The slope became steeper lower down the hill, and the regular ledges then reappeared. Another of my sons observed, on the inland side of Beachy Head, where the surface sloped at about 25°, many short little embankments like those just mentioned. They extended horizontally and were from a few inches to two or three feet in length. They supported tufts of grass growing vigorously. The average thickness of the mould of which they were formed, taken from nine measurements, was 4·5 inches; while that of the mould above and beneath them was on an average only 3·2 inches, and on each side, on the same level, 3·1 inches. On the upper parts of the slope, these embankments showed no signs of having been trampled on by sheep, but in the lower parts such signs were fairly plain. No long continuous ledges had here been formed.

If the little embankments above the Corniche road, which Dr. King saw in the act of formation by the accumulation of disintegrated and rolled worm-castings, were to become confluent along horizontal lines, ledges would be formed. Each embankment would tend to extend laterally by the lateral extension of the arrested castings; and animals grazing on a steep slope would almost certainly make use of every prominence at nearly the same level, and would indent the turf between them; and such intermediate indentations would again arrest the castings. An irregular ledge when once formed would also tend to become more regular and horizontal by some of the castings rolling laterally from the higher to the lower parts, which would thus be raised. Any projection beneath a ledge would not afterwards receive distintegrated matter from above, and would tend to be obliterated by rain and other atmospheric agencies. There is some analogy between the formation, as here supposed, of these ledges, and that of the ripples of wind-drifted sand as described by Lyell.*

The steep, grass-covered sides of a mountainous valley in Westmoreland, called Grisedale, was marked in many places with innumerable, almost horizontal, little ledges, or rather lines of miniature cliffs. Their formation was in no way connected with the action of worms, for castings could not anywhere be seen (and their absence is an inexplicable fact) although the turf lay in many places over a considerable thickness of

* 'Elements of Geology,' 1865, p. 20.

boulder-clay and moraine rubbish. Nor, as far as I could judge, was the formation of these little cliffs at all closely connected with the trampling of cows or sheep. It appeared as if the whole superficial, somewhat argillaceous earth, while partially held together by the roots of the grasses, had slided a little way down the mountain sides; and in thus sliding, had yielded and cracked in horizontal lines, transversely to the slope.

Castings blown to leeward by the wind.—We have seen that moist castings flow, and that disintegrated castings roll down any inclined surface; and we shall now see that castings, recently ejected on level grass-covered surfaces, are blown during gales of wind accompanied by rain to leeward. This has been observed by me many times on many fields during several successive years. After such gales, the castings present a gently inclined and smooth, or sometimes furrowed, surface to windward, while they are steeply inclined or precipitous to leeward, so that they resemble on a miniature scale glacier-ground hillocks of rock. They are often cavernous on the leeward side, from the upper part having curled over the lower part. During one unusually heavy south-west gale with torrents of rain, many castings were wholly blown to leeward, so that the mouths of the burrows were left naked and exposed on the windward side. Recent castings naturally flow down an inclined surface, but on a grassy field, which sloped between 10° and 15°, several were found after a heavy gale blown up the slope. This likewise occurred on another occasion on a part of my lawn where the slope was somewhat less. On a third occasion, the castings on the steep, grass-covered sides of a valley, down which a gale had blown, were directed obliquely instead of straight down the slope; and this was obviously due to the combined action of the wind and gravity. Four castings on my lawn, where the downward inclination was 0° 45′, 1°, 3° and 3° 30′ (mean 1° 49′) towards the north-east, after a heavy south-west gale with rain, were divided across the mouths of the burrows and weighed in the manner formerly described. The mean weight of the earth below the mouths of burrows and to leeward, was to that above the mouths and on the windward side as $2\frac{3}{4}$ to 1; whereas we have seen that with several castings which had flowed down slopes having a mean inclination of 9° 26′, and with three castings where the inclination was above 12°, the proportional weight of the earth below to that above the burrows was as only 2 to 1. These several cases show how efficiently gales of wind accompanied by rain act in displacing recently-ejected castings. We may therefore conclude that even a moderately strong wind will produce some slight effect on them.

Dry and indurated castings, after their disintegration into small fragments or pellets, are sometimes, probably often, blown by a strong wind to leeward. This was observed on four occasions, but I did not sufficiently attend to this point. One old casting on a gently sloping bank was blown quite away by a strong south-west wind. Dr. King believes that the wind removes the greater part of the old crumbling castings near Nice. Several old castings on my lawn were marked with pins and protected from any disturbance. They were examined after an interval of 10 weeks, during which time the weather had been alternately dry and rainy. Some, which were of a yellowish colour had been washed almost completely away, as could be seen by the colour of the surrounding ground. Others had completely disappeared, and these no doubt had been blown away. Lastly, others still remained and would long remain, as blades of grass had grown through them. On poor pasture land, which has never been rolled and has not been much trampled on by animals, the whole surface is sometimes dotted with little pimples, through and on which grass grows; and these pimples consist of old worm-castings.

In all the many observed cases of soft castings blown to leeward, this had been effected by strong winds accompanied by rain. As such winds in England generally blow from the south and south-west, earth must on the whole tend to travel over our fields in a north and north-east direction. This fact is interesting, because it might be thought that none could be removed from a level, grass-covered surface by any means. In thick and level woods, protected from the wind, castings will never be removed as long as the wood lasts; and mould will here tend to accumulate to the depth at which worms can work. I tried to procure evidence as to how much mould is blown, whilst in the state of castings, by our wet southern gales to the north-east, over open and flat land, by looking to the level of the surface on opposite sides of old trees and hedge-rows; but I failed owing to the unequal growth of the roots of trees and to most pasture-land having been formerly ploughed.

On an open plain near Stonehenge, there exist shallow circular trenches, with a low embankment outside, surrounding level spaces 50 yards in diameter. These rings appear very ancient, and are believed to be contemporaneous with the Druidical stones. Castings ejected within these circular spaces, if blown to the north-east by south-west winds would form a layer of mould within the trench, thicker on the north-eastern than on any other side. But the site was not favourable for the action of worms, for the mould over the surrounding Chalk formation with flints, was only 3·37 inches in thickness, from a mean of six observations made at a distance of 10 yards outside the embankment. The thickness of the mould within two of the circular trenches was measured every 5 yards all round, on the inner sides near the bottom. My son Horace protracted these measurements on paper; and though the curved line representing the thickness of the mould was extremely irregular, yet in both diagrams it could be seen to be thicker on the north-eastern side than elsewhere. When a mean of all the measurements in both the trenches was laid down and the line smoothed, it was obvious that the mould was thickest in the quarter of the circle between north-west and north-east; and thinnest in the quarter between south-east and south-west, especially at this latter point. Besides the foregoing measurements, six others were taken near together in one of the circular trenches, on the north-east side; and the mould here had a mean thickness of 2·29 inches; while the mean of six other measurements on the south-west side was only 1·46 inches. These observations indicate that the castings had been blown by the south-west winds from the circular enclosed space into the trench on the north-east side; but many more measurements in other analogous cases would be requisite for a trustworthy result.

The amount of fine earth brought to the surface under the form of castings, and afterwards transported by the winds accompanied by rain, or that which flows and rolls down an inclined surface, no doubt is small in the course of a few scores of years; for otherwise all the inequalities in our pasture fields would be smoothed within a much shorter period than appears to be the case. But the amount which is thus transported in the course of thousands of years cannot fail to be considerable and deserves attention. É. de Beaumont looks at the vegetable mould which everywhere covers the land as a fixed line or zero, from which the amount of denudation may be measured.* He ignores the continued formation of fresh mould by the disintegration of the underlying rocks and fragments of rock; and it is curious to find how much more philosophical were the views, main-

* "Leçons de Géologie pratique, 1845; cinquième Leçon.' All Élie de Beaumont's arguments are admirably controverted by Prof. A. Geikie in his essay in Transact. Geolog. Soc. of Glasgow, vol. iii. p. 153, 1868.

tained long ago, by Playfair, who, in 1802, wrote, "in the permanence of a coat of "vegetable mould on the surface of the earth, "we have a demonstrative proof of the con-"tinued destruction of the rocks."*

Ancient encampments and tumuli.—É. de Beaumont adduces the present state of many ancient encampments and tumuli and of old ploughed fields, as evidence that the surface of the land undergoes hardly any degradation. But it does not appear that he ever examined the thickness of the mould over different parts of such old remains. He relies chiefly on indirect, but apparently trustworthy, evidence that the slopes of the old embankments are the same as they originally were; and it is obvious that he could know nothing about their original heights. In Knole Park a mound had been thrown up behind the rifle-targets, which appeared to have been formed of earth originally supported by square blocks of turf. The sides sloped, as nearly as I could estimate them, at an angle of 45° or 50° with the horizon, and they were covered, especially on the northern side, with long coarse grass,

* 'Illustrations of the Huttonian Theory of the Earth,' p. 107.

beneath which many worm-castings were found. These had flowed bodily downwards, and others had rolled down as pellets. Hence it is certain that as long as a mound of this kind is tenanted by worms, its height will be continually lowered. The fine earth which flows or rolls down the sides of such a mound accumulates at its base in the form of a talus. A bed, even a very thin bed, of fine earth is eminently favourable for worms; so that a greater number of castings would tend to be ejected on a talus thus formed than elsewhere; and these would be partially washed away by every heavy shower and be spread over the adjoining level ground. The final result would be the lowering of the whole mound, whilst the inclination of the sides would not be greatly lessened. The same result would assuredly follow with ancient embankments and tumuli; except where they had been formed of gravel or of nearly pure sand, as such matter is unfavourable for worms. Many old fortifications and tumuli are believed to be at least 2000 years old; and we should bear in mind that in many places about one inch of mould is brought to the surface in 5 years or two inches in 10 years. Therefore in so long a period as 2000 years, a large amount of earth will have been repeatedly brought to the surface on most old embankments and tumuli, especially on the talus round their bases, and much of this earth will have been washed completely away. We may therefore conclude that all ancient mounds, when not formed of materials unfavourable to worms, will have been somewhat lowered in the course of centuries, although their inclinations may not have been greatly changed.

Fields formerly ploughed.—From a very remote period and in many countries, land has been ploughed, so that convex beds, called crowns or ridges, usually about 8 feet across and separated by furrows, have been thrown up. The furrows are directed so as to carry off the surface water. In my attempts to ascertain how long a time these crowns and furrows last, when ploughed land has been converted into pasture, obstacles of many kinds were encountered. It is rarely known when a field was last ploughed; and some fields which were thought to have been in pasture from time immemorial were afterwards discovered to have been ploughed only 50 or 60 years before. During the early part of the present century, when the price of corn was very high, land of all kinds seems to have been ploughed in Britain. There is, however, no reason to doubt that in many cases the old crowns and furrows have been preserved from a very ancient period.* That they should have been preserved for very unequal lengths of time would naturally follow from the crowns, when first thrown up, having differed much in height in different districts, as is now the case with recently ploughed land.

In old pasture fields, the mould, wherever measurements were made, was found to be from ½ to 2 inches thicker in the furrows than

* Mr. E. Tylor in his Presidential address ('Journal of the Anthropological Institute,' May 1880, p. 451) remarks: "It appears from several papers of the Berlin Society as to the German 'high-fields' or 'heathen-fields' (Hochäcker, and Heidenäcker) that they correspond much in their situation on hills and wastes with the 'elf-furrows' of Scotland, which popular mythology accounts for by the story of the fields having been put under a Papal interdict, so that people took to cultivating the hills. There seems reason to suppose that, like the tilled plots in the Swedish forests which tradition ascribes to the old 'hackers,' the German heathen-fields represent tillage by an ancient and barbaric population."

on the crowns; but this would naturally follow from the finer earth having been washed from the crowns into the furrows before the land was well clothed with turf; and it is impossible to tell what part worms may have played in the work. Nevertheless from what we have seen, castings would certainly tend to flow and to be washed during heavy rain from the crowns into the furrows. But as soon as a bed of fine earth had by any means been accumulated in the furrows, it would be more favourable for worms than the other parts, and a greater number of castings would be thrown up here than elsewhere; and as the furrows on sloping land are usually directed so as to carry off the surface water, some of the finest earth would be washed from the castings which had been here ejected and be carried completely away. The result would be that the furrows would be filled up very slowly, while the crowns would be lowered perhaps still more slowly by the flowing and rolling of the castings down their gentle inclinations into the furrows.

Nevertheless it might be expected that old furrows, especially those on a sloping surface, would in the course of time be filled up and disappear. Some careful observers, however, who examined fields for me in Gloucestershire and Staffordshire, could not detect any difference in the state of the furrows in the upper and lower parts of sloping fields, supposed to have been long in pasture; and they came to the conclusion that the crowns and furrows would last for an almost endless number of centuries. On the other hand the process of obliteration seems to have commenced in some places. Thus in a grass field in North Wales, known to have been ploughed about 65 years ago, which sloped at an angle of 15° to the north-east, the depth of the furrows (only 7 feet apart) was carefully measured, and was found to be about 4½ inches in the upper part of the slope, and only 1 inch near the base, where they could be traced with difficulty. On another field sloping at about the same angle to the south-west, the furrows were scarcely perceptible in the lower part; although these same furrows when followed on to some adjoining level ground were from 2½ to 3½ inches in depth. A third and closely similar case was observed. In a fourth case, the mould in a furrow in the upper part of a sloping field was 2½ inches, and in the lower part 4½ inches in thickness.

On the Chalk Downs at about a mile distance from Stonehenge, my son William examined a grass-covered, furrowed surface, sloping at from 8° to 10°, which an old shepherd said had not been ploughed within the memory of man. The depth of one furrow was measured at 16 points in a length of 68 paces, and was found to be deeper where the slope was greatest and where less earth would naturally tend to accumulate, and at the base it almost disappeared. The thickness of the mould in this furrow in the upper part was 2½ inches, which increased to 5 inches a little above the steepest part of the slope; and at the base, in the middle of the narrow valley, at a point which the furrow if continued would have struck, it amounted to 7 inches. On the opposite side of the valley, there were very faint, almost obliterated, traces of furrows. Another analogous but not so decided a case was observed at a few miles distance from Stonehenge. On the whole it appears that the crowns and furrows on land formerly ploughed, but now covered with grass, tend slowly to disappear when the surface is inclined; and this is probably in large part due to the action of worms; but that the crowns and furrows last for a very long time when the surface is nearly level.

Formation and amount of mould over the Chalk Formation.—Worm-castings are often ejected in extraordinary numbers on steep, grass-covered slopes, where the Chalk comes close to the surface, as my son William observed near Winchester and elsewhere. If such castings are largely washed away during heavy rains, it is difficult to understand at first how any mould can still remain on our Downs, as there does not appear any evident means for supplying the loss. There is, moreover, another cause of loss, namely in the percolation of the finer particles of earth into the fissures in the chalk and into the chalk itself. These considerations led me to doubt for a time whether I had not exaggerated the amount of fine earth which flows or rolls down grass-covered slopes under the form of castings; and

I sought for additional information. In some places, the castings on Chalk Downs consist largely of calcareous matter, and here the supply is of course unlimited. But in other places, for instance on a part of Teg Down near Winchester, the castings were all black and did not effervesce with acids. The mould over the chalk was here only from 3 to 4 inches in thickness. So again on the plain near Stonehenge, the mould, apparently free from calcareous matter, averaged rather less than 3½ inches in thickness. Why worms should penetrate and bring up chalk in some places and not in others I do not know.

In many districts where the land is nearly level, a bed several feet in thickness of red clay full of unworn flints overlies the Upper Chalk. This overlying matter, the surface of which has been converted into mould, consists of the undissolved residue from the chalk. It may be well here to recall the case of the fragments of chalk buried beneath worm-castings on one of my fields, the angles of which were so completely rounded in the course of 29 years that the fragments now resembled water-worn pebbles. This must have been effected by the carbonic acid in the rain and in the ground, by the humus-acids, and by the corroding power of living roots. Why a thick mass of residue has not been left on the Chalk, wherever the land is nearly level, may perhaps be accounted for by the percolation of the fine particles into the fissures, which are often present in the chalk and are either open or are filled up with impure chalk, or into the solid chalk itself. That such percolation occurs can hardly be doubted. My son collected some powdered and fragmentary chalk beneath the turf near Winchester; the former was found by Colonel Parsons, R.E., to contain 10 per cent., and the fragments 8 per cent. of earthy matter. On the flanks of the escarpment near Abinger in Surrey, some chalk close beneath a layer of flints, 2 inches in thickness and covered by 8 inches of mould, yielded a residue of 3·7 per cent. of earthy matter. On the other hand the Upper Chalk properly contains, as I was informed by the late David Forbes who had made many analyses, only from 1 to 2 per cent. of earthy matter; and two samples from pits near my house contained 1·3 and 0·6 per cent. I mention these latter cases because, from the thickness of the overlying bed of red clay with flints, I had imagined that the underlying chalk might here be less pure than elsewhere. The cause of the residue accumulating more in some places than in others, may be attributed to a layer of argillaceous matter having been left at an early period on the chalk, and this would check the subsequent percolation of earthy matter into it.

From the facts now given we may conclude that castings ejected on our Chalk Downs suffer some loss by the percolation of their finer matter into the chalk. But such impure superficial chalk, when dissolved, would leave a larger supply of earthy matter to be added to the mould than in the case of pure chalk. Besides the loss caused by percolation, some fine earth is certainly washed down the sloping grass-covered surfaces of our Downs. The washing-down process, however, will be checked in the course of time; for although I do not know how thin a layer of mould suffices to support worms, yet a limit must at last be reached; and then their castings would cease to be ejected or would become scanty.

The following cases show that a considerable quantity of fine earth is washed down. The thickness of the mould was measured at points 12 yards apart across a small valley in the Chalk near Winchester. The sides sloped gently at first; then became inclined at about 20°; then more gently to near the bottom, which transversely was almost level and about 50 yards across. In the bottom, the mean thickness of the mould from five measurements was 8·3 inches; whilst on the sides of the valley, where the inclination varied between 14° and 20°, its mean thickness was rather less than 3·5 inches. As the turf-covered bottom of the valley sloped at an angle of only between 2° and 3°, it is probable that most of the 8·3-inch layer of mould had been washed down from the flanks of the valley, and not from the upper part. But as a shepherd said that he had seen water flowing in this valley after the sudden thawing of snow, it is possible that some earth may have been brought down from the upper part; or, on the other hand, that some may have been carried further down the valley. Closely similar results, with respect to the thickness of the mould, were obtained in a neighbouring valley.

St. Catherine's Hill, near Winchester, is 327 feet in height, and consists of a steep cone of chalk about ¼ of a mile in diameter. The upper part was converted by the Romans, or, as some think, by the ancient Britons, into an encampment, by the excavation of a deep and broad ditch all round it. Most of the chalk removed during the work was thrown upwards, by which a projecting bank was formed; and this effectually prevents worm-castings (which are numerous in parts), stones, and other objects from being washed or rolled into the ditch. The mould on the upper and fortified part of the hill was found to be in most places only from 2½ to 3½ inches in thickness; whereas it had accumulated at the foot of the embankment above the ditch to a thickness in most places of from 8 to 9½ inches. On the embankment itself the mould was only 1 to 1½ inch in thickness; and within the ditch at the bottom it varied from 2½ to 3½, but was in one spot 6 inches in thickness. On the north-west side of the hill, either no embankment had ever been thrown up above the ditch, or it had subsequently been removed; so that here there was nothing to prevent worm-castings, earth and stones being washed into the ditch, at the bottom of which the mould formed a layer from 11 to 22 inches in thickness. It should however be stated that here and on other parts of the slope, the bed of mould often contained fragments of chalk and flint which had obviously rolled down at different times from above. The interstices in the underlying fragmentary chalk were also filled up with mould.

My son examined the surface of this hill to its base in a south-west direction. Beneath the great ditch, where the slope was about 24°, the mould was very thin, namely from $1\frac{1}{2}$ to $2\frac{1}{2}$ inches; whilst near the base, where the slope was only 3° to 4°, it increased to between 8 and 9 inches in thickness. We may therefore conclude that on this artificially modified hill, as well as in the natural valleys of the neighbouring Chalk Downs, some fine earth, probably derived in large part from worm-castings, is washed down, and accumulates in the lower parts, notwithstanding the percolation of an unknown quantity into the underlying chalk; a supply of fresh earthy matter being afforded by the dissolution of the chalk through atmospheric and other agencies.

CHAPTER VII.

CONCLUSION.

Summary of the part which worms have played in the history of the world—Their aid in the disintegration of rocks—In the denudation of the land—In the preservation of ancient remains—In the preparation of the soil for the growth of plants—Mental powers of worms—Conclusion.

WORMS have played a more important part in the history of the world than most persons would at first suppose. In almost all humid countries they are extraordinarily numerous, and for their size possess great muscular power. In many parts of England a weight of more than ten tons (10,516 kilogrammes) of dry earth annually passes through their bodies and is brought to the surface on each acre of land; so that the whole superficial bed of vegetable mould passes through their bodies in the course of every few years. From the collapsing of the old burrows the mould is in constant though slow movement, and the particles composing it are thus rubbed together. By these means fresh surfaces are continually exposed to the action of the carbonic acid in the soil, and of the humus-acids which appear to be still more efficient in the decomposition of rocks. The generation of the humus-acids is probably hastened during the digestion of the many half-decayed leaves which worms consume. Thus the particles of earth, forming the superficial mould, are subjected to conditions eminently favourable for their decomposition and disintegration. Moreover, the particles of the softer rocks suffer some amount of mechanical trituration in the muscular gizzards of worms, in which small stones serve as mill-stones.

The finely levigated castings, when brought to the surface in a moist condition, flow during rainy weather down any moderate slope; and the smaller particles are washed far down even a gently inclined surface. Castings when dry often crumble into small pellets and these are apt to roll down any sloping surface. Where the land is quite level and is covered with herbage, and where the climate is humid so that much dust cannot be blown away, it appears at first sight impossible that there should be any appreciable amount of subaerial denudation; but worm-castings are blown, especially whilst moist and viscid, in one uniform direction by the prevalent winds which are accompanied by rain. By these several means the superficial mould is prevented from accumulating to a great thickness; and a thick bed of mould checks in many ways the disintegration of the underlying rocks and fragments of rock.

The removal of worm castings by the above means leads to results which are far from insignificant. It has been shown that a layer of earth, ·2 of an inch in thickness, is in many places annually brought to the surface per acre; and if a small part of this amount flows, or rolls, or is washed, even for a short distance down every inclined surface, or is repeatedly blown in one direction, a great effect will be produced in the course of ages. It was found by measurements and calculations that on a surface with a mean inclination of 9° 26′, 2·4 cubic inches of earth which had been ejected by worms crossed, in the course of a year, a horizontal line one yard in length; so that 240 cubic inches would cross a line 100 yards in length. This latter amount in a damp state would weigh 11½ pounds. Thus a considerable weight of earth is continually moving down each side of every valley, and will in time reach its bed. Finally this earth will be transported by the streams flowing in the valleys into the ocean, the great receptacle for all matter denuded from the land. It is known from the amount of sediment annually delivered into the sea by the Mississippi, that its enormous drainage-area must on an average be lowered ·00263 of an inch each year; and this would suffice in four and half million years to lower the whole drainage-area to the level of the sea-shore. So that, if a small fraction of the layer of fine earth, ·2 of an inch in thickness, which is annually brought to the surface by worms, is carried away, a great result cannot fail to be produced within a period which no geologist considers extremely long.

Archæologists ought to be grateful to worms, as they protect and preserve for an indefinitely long period every object, not liable to decay, which is dropped on the surface of the land, by burying it beneath their castings. Thus, also, many elegant and curious tesselated pavements and other ancient remains have been preserved; though no doubt the worms have in these cases been largely aided by earth washed and blown from the adjoining land, especially when cultivated. The old tesselated pavements have, however, often suffered by having subsided unequally from being unequally undermined by the worms. Even old massive walls may be undermined and subside; and no building is in this respect safe, unless the foundations lie 6 or 7 feet beneath the surface, at a depth at which worms cannot work. It is probable that many monoliths and some old walls have fallen down from having been undermined by worms.

Worms prepare the ground in an excellent manner for the growth of fibrous-rooted plants and for seedlings of all kinds. They periodically expose the mould to the air, and sift it so that no stones larger than the particles which they can swallow are left in it. They mingle the whole intimately together, like a gardener who prepares fine soil for his choicest plants. In this state it is well fitted to retain moisture and to absorb all soluble substances, as well as for the process of nitrification. The bones of dead animals, the harder parts of insects, the shells of land-molluscs, leaves, twigs, &c., are before long all buried beneath the accumulated castings of worms, and are thus brought in a more or less decayed state within reach of the roots of plants. Worms likewise drag an infinite number of dead leaves and other parts of plants into their burrows, partly for the sake of plugging them up and partly as food.

The leaves which are dragged into the burrows as food, after being torn into the finest shreds, partially digested, and saturated with the intestinal and urinary secretions, are commingled with much earth. This earth forms the dark coloured, rich humus which almost everywhere covers the surface of the land with a fairly well-defined layer or mantle. Von Hensen* placed two worms in a vessel

* 'Zeitschrift für wissenschaft. Zoolog.' B. xxviii. 1877, p. 360.

18 inches in diameter, which was filled with sand, on which fallen leaves were strewed; and these were soon dragged into their burrows to a depth of 3 inches. After about 6 weeks an almost uniform layer of sand, a centimeter (·4 inch) in thickness, was converted into humus by having passed through the alimentary canals of these two worms. It is believed by some persons that worm-burrows, which often penetrate the ground almost perpendicularly to a depth of 5 or 6 feet, materially aid in its drainage; notwithstanding that the viscid castings piled over the mouths of the burrows prevent or check the rain-water directly entering them. They allow the air to penetrate deeply into the ground. They also greatly facilitate the downward passage of roots of moderate size; and these will be nourished by the humus with which the burrows are lined. Many seeds owe their germination to having been covered by castings; and others buried to a considerable depth beneath accumulated castings lie dormant, until at some future time they are accidentally uncovered and germinate.

Worms are poorly provided with sense-organs, for they cannot be said to see, although they can just distinguish between light and darkness; they are completely deaf, and have only a feeble power of smell; the sense of touch alone is well developed. They can therefore learn little about the outside world, and it is surprising that they should exhibit some skill in lining their burrows with their castings and with leaves, and in the case of some species in piling up their castings into tower-like constructions. But it is far more surprising that they should apparently exhibit some degree of intelligence instead of a mere blind instinctive impulse, in their manner of plugging up the mouths of their burrows. They act in nearly the same manner as would a man, who had to close a cylindrical tube with different kinds of leaves, petioles, triangles of paper, &c., for they commonly seize such objects by their pointed ends. But with thin objects a certain number are drawn in by their broader ends. They do not act in the same unvarying manner in all cases, as do most of the lower animals; for instance, they do not drag in leaves by their foot-stalks, unless the basal part of the blade is as narrow as the apex, or narrower than it.

When we behold a wide, turf-covered expanse, we should remember that its smoothness, on which so much of its beauty depends, is mainly due to all the inequalities having been slowly levelled by worms. It is a marvellous reflection that the whole of the superficial mould over any such expanse has passed, and will again pass, every few years through the bodies of worms. The plough is one of the most ancient and most valuable of man's inventions; but long before he existed the land was in fact regularly ploughed, and still continues to be thus ploughed by earth-worms. It may be doubted whether there are many other animals which have played so important a part in the history of the world, as have these lowly organised creatures. Some other animals, however, still more lowly organised, namely corals, have done far more conspicuous work in having constructed innumerable reefs and islands in the great oceans; but these are almost confined to the tropical zones.

INDEX.

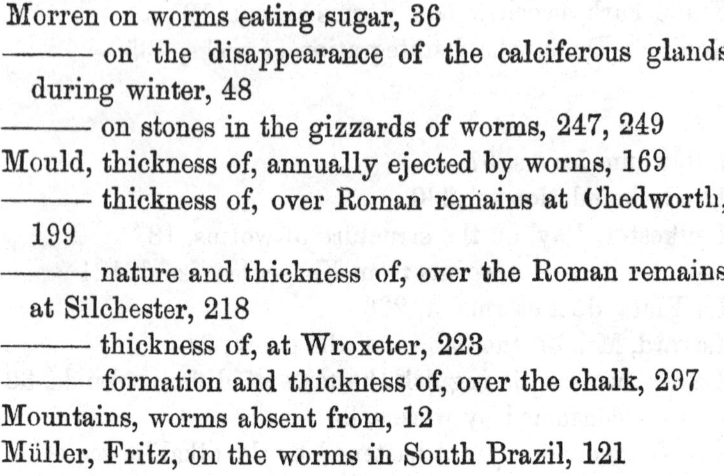

Printed in the United States
By Bookmasters